To my parents
For my son

Jan A. Pechenik

To my past, present
and future students

Bernard C. Lamb

How to Write about Biology

JAN A. PECHENIK
Tufts University

BERNARD C. LAMB
Imperial College, London University

Adaptation published by arrangement with HarperCollins Publishers, Inc, USA

This edition first published in 1994 by
HarperCollins College Division
An imprint of HarperCollins Publishers Ltd, UK
77–85 Fulham Palace Road
Hammersmith
London W6 8JB
Reprinted 1995
Reprinted 1996 (Addison Wesley Longman)

British Library Cataloguing in Publication Data
A catalogue record for this book is available from the British Library.

ISBN 0582-30922-0

Typeset by Dorchester Typesetting Group Ltd
Printed and bound by Bookcraft (Bath) Ltd
Cover design: Patrick Knowles
Cover illustration: human hair

Contents

Preface

When I read the second edition of the American forerunner of this book, *A Short Guide to Writing about Biology*, by J.A. Pechenik, I was delighted to find such a clear and helpful guide for students. There was also much advice that I could immediately apply to improve my own writing, even though I had had over 20 years of experience as a professional biologist, university teacher and research geneticist. As it was written for students at American universities, some of it was not applicable to British and European students, so I was very pleased to be asked to prepare a new version for British and European students, incorporating a lot of new material of my own.

Although this book is largely addressed to undergraduates and post-graduates in biology, it is also very suitable for those working in related areas such as agriculture, medicine, biochemistry and biotechnology, and for professional biologists.

Some of the changes from the American forerunner of this book are just adaptations for the UK and Europe; others are updatings, especially on computer-searching of literature, where it is a pleasure to acknow-ledge the help of Mrs Janice Yeadon of Imperial College, London. There are new sections on *Laboratory and Field Drawings*, on *Answering Exam Questions*, on *Preparing Research Posters*, and on *SI Units and Prefixes.*

The major new chapter is on *Improving Punctuation, Word-Choice, Spelling and Grammar.* This incorporates results of my research into the English standards of biologists and of students in other sciences, arts, engineering and medicine (*e.g. A National Survey of UK Undergraduates' Standards of English*, B.C. Lamb, 1992, The Queen's English Society, London). It has also been guided by other people's comments (from academics, employers, industrialists and teachers) on the English of students, graduates and job applicants. Many faults in the writing of biologists relate to English, rather than to biology, so the most relevant parts of basic English are included in a biological context, with exercises. Some students' problems with spelling, punctuation, word-choice and sentence structure seriously handicap them in expressing themselves clearly, accurately and coherently in their chosen subject, yet the solutions and rules are often very simple.

I have added more short exercises, with answers provided, to assist learning by giving the reader practice in usage, and in spotting and correcting faults. Most examples are drawn from writing by students or colleagues; some are from other sources such as scientific journals and newspapers. I am grateful to some of my colleagues for sending me examples from their own areas of biology, and for their advice on what to include in this book. Any comments, criticisms, suggestions and advice from users of this book, whether students or professional biologists, will be received gratefully, and should be sent to Dr B.C. Lamb, Biology Department, Imperial College, London, SW7 2BB, UK.

There were problems in changing a single-authored book to a co-authored one, as the style was very personal, with lots of 'I' references and phrases such as 'my wife', where substituting 'we' and 'our' was undesirable. I have therefore added the initials '(JP)' or '(BL)' after 'I' or 'my', where appropriate.

Finally, I wish to thank all those who have helped with advice or with checking this book, especially Dr Meriel Jones, Dr Simon Archer, Dr Adrian Evans, Dr Valerie Brown, Dr John Brady, Mrs Brenda Lamb and Mrs Jennifer Chew.

Dr Bernard C. Lamb
London, January 1994

1. Introduction and General Rules

The clear and precise communication of ideas and facts through writing is among the most important skills that can be mastered by students. Effective writing in biology requires time, effort, clear examples, instruction and practice. It also needs correction by more experienced writers, leading to self-correction later. In biology classes there is often much writing to do but little time to concentrate on basic skills. Developing your writing skills is worth every bit of effort it takes, as it helps you with lab reports, essays, examinations, theses, job applications and private correspondence.

Writing in biology is closely connected with doing, understanding and learning biology, so this book sometimes goes beyond purely writing aspects. For example, it includes advice on giving talks and on literature-searching. This book gives technical advice on how to write essays, lab reports, exam answers and other things which undergraduate and post-graduate biologists have to write. Because a major problem in the writing of many biologists is a lack of mastery of elements of the English language, such as spelling, punctuation, word-choice and grammar, those aspects are also covered, in a biological context.

WHAT DO BIOLOGISTS WRITE, WHEN AND WHY?

Undergraduates mainly write lecture notes, tutorial essays, laboratory reports of practical experiments, and examination answers. As you progress through your undergraduate years, you may also have to write longer essays (sometimes called dissertations or extended essays) based on an original literature search, or write criticisms or summaries of scientific papers, and eventually a thesis based on an individual research project. Other important items to write will be a Curriculum Vitae (CV), to summarise your qualifications and achievements for potential employers, and letters of application for jobs, research studentships or places on taught post-graduate courses. If you apply for vacation jobs, you will need those job-application writing skills very early on.

Final-year undergraduates are sometimes asked to write draft scientific papers, as useful exercises for their later careers. Especially in their final year, undergraduates often need to prepare talks to give to their fellow students, and talks need very careful planning of slides, overhead transparencies, blackboard diagrams and other written materials.

Like undergraduates, post-graduates taking taught courses will also write essays and lab reports, but at a more advanced level, with a greater need for literature-searching skills. Those doing research projects must be able to write an extended and well-organised thesis on their research, with a lot of background literature-searching. They must also be able to write scientific papers. Post-graduates often have to give talks on their own work or that of others, to fellow post-graduates or to research meetings, such as those of national societies in their subject area.

Professional biologists will usually write scientific papers, review articles, research proposals and grant applications, reports for clients or employers, memoranda, minutes of meetings, assessments of manuscripts and of grant proposals written by other biologists, and letters to the head of department, to editors and to actual or potential collaborators. Teachers of biology in schools and universities may also have to write syllabuses, lectures, class practical schedules, examination papers and references for students.

All biologists, whether students or professionals, have to prepare facts, arguments and suggestions, for communication to others, orally and in writing. Like a good essay, a lecture presents information in an orderly manner; to research and write a coherent, well-planned lecture can easily take a whole day. University teachers and other research workers also write grant proposals in the hope of obtaining funding for research. Even well-written proposals have a difficult time; poorly written proposals generally don't stand a chance.

After doing research, we have to prepare the results for publication. Unless they are purely theoretical, these papers are lab or field reports based on data collected over a much longer period of time than the typical student lab session. The goal is to present data clearly and to interpret them thoroughly and convincingly in the context of other work and basic biological principles. Preparing of research papers typically involves the following steps:

- Organising the data
- Preparing a first draft, following the procedures outlined in Chapters 3 and 9
- Revising and reprinting (or retyping) the paper
- Asking one or more colleagues to read the paper critically
- Revising the paper in accordance with the readers' comments
- Reprinting and proof-reading the paper
- Sending the paper to the editor of the chosen journal

The editor sends the manuscript to be reviewed by two or more biologists. Their comments, along with those of the editor, are then sent to the author, who must often rewrite the paper, sometimes extensively. The editor may then accept or reject the revised manuscript, or may request that it be rewritten again.

Biologists obviously write about biology, but they also write about other things, such as students; letters of recommendation are especially challenging for staff because they are so important to you. They must be written clearly, developed logically and proof-read carefully if they are to argue convincingly on your behalf. Then there are progress reports, committee reports, letters and memoranda. All this writing involves thinking, organising, nailing down a convincing argument on paper, revising, retyping and proof-reading.

We hope you are now convinced that effective writing is highly relevant in a scientific career. When biology students receive criticisms of their writing, they often complain that 'This is not an English course.' These students do not understand that clear, correct, concise, logical writing is a vital tool of the biologist's trade, and that learning how to write well is more important than learning how to use a balance, extract a protein, measure a nerve impulse or run an electrophoretic gel. Unlike these specialised lab techniques, mastering effective writing will reward you regardless of the field in which you eventually find yourself. For example, the difference between a well-crafted and a poorly-crafted letter of application is often the difference between getting the job or losing to another contender.

SOME KEYS TO SUCCESS

All good writing involves two struggles: the struggle for understanding and the struggle to communicate that understanding to a reader. Like the making of omelettes, the skill improves with practice, and being aware of certain key principles will ease the way considerably. These rules are discussed more fully in later chapters.

1. **Work to understand your sources.** When writing lab reports, spend time wrestling with your data until you are convinced you see the significance of what you have done. If appropriate, try graphing your results in different ways. When taking notes from books or research articles, re-read sentences you don't understand and look up words that puzzle you. Too few students take this struggle for understanding seriously enough, but all good scientific writing begins with this first struggle. You can excel – now, and later in your career – by being one of the few who meet this challenge head-on. If you don't commit yourself to winning the struggle for understanding, you will end up with nothing to say or what you say will be wrong.

2. **Think of an outline plan of the piece before you begin to write.** Much of the real work of writing is in the thinking that must precede each draft. Effective writing is like effective sailing: you must take the time to plot your course before getting too far from port. Your ideas about where you are going and how best to get there may change as you write your paper, since the act of writing clarifies your thinking

and often brings entirely new ideas. Nevertheless, you must have some plan in mind when you begin to write: this plan evolves from thoughtful consideration of your notes. Think first, then write, then revise.

3. **Write to illuminate, not to impress.** Use the simplest words and the simplest phrasing consistent with that goal. Define all specialised terminology. In general, if a term was new to you, it should be defined in your writing. Don't try to impress the reader with big words and a technical vocabulary: focus on getting your point across.

4. **Make a statement and back it up.** Remember, you are making an argument. In any argument, a statement of fact or opinion becomes convincing to the critical reader only when supported by evidence or explanation: provide it. You might, for instance, write, 'Among the vertebrates, the development of sperm is triggered by the release of the hormone testosterone (Browder, 1984).' In this case, the statement is supported by reference to a book written by L. Browder in 1984. In the following example, a statement is backed up by reference to the writer's own data: 'Some wavelengths of light were more effective than others in promoting photosynthesis. For example, the rate of oxygen production at 650 nm was nearly four times greater than that recorded for the same plants when using a wavelength of 550 nm (Figure 2).' Advice on citing literature is given in Chapter 3, page 90. In ordinary tutorial essays, it is not necessary to give references for each statement, but references are much more important in formal theses and scientific papers. Explain any statements that are not clear or self-evident. Long-established facts, such as: 'Aspergillus is a common filamentous fungus', require no supporting references, even in theses.

   References to papers or books written by two authors must include the names of both authors (*e.g.* Burns and Allen, 1946). When there are more than two authors, only the first author's name is written out (*e.g.* Fried *et al.*, 1990). An author's first name is never included in the citation. A statement made by your lecturer should be cited as a personal communication (*e.g.* 'R. A. Fisher, personal communication'). Refer to a lab manual or schedule by its author (*e.g.* Chase, 1992), or as follows: (Second Year Molecular Biology Laboratory Manual, 1993). 'See attached schedule' is often adequate for informal lab reports.

5. **Always distinguish fact from possibility.** In the course of examining your data or reading your notes, you may form an opinion, but be careful not to state your opinion as though it were fact. 'Species *X* lacks the ability to respond to sucrose' is a statement of fact and should usually be supported with a reference. 'Our data suggest that species *X* lacks the ability to respond to sucrose' or 'Species *X* seems unable to respond to sucrose' expresses your opinion and should be supported by drawing the reader's attention to key elements of your data.

6. **Write exactly what you mean.** Words are tricky; if they don't end up in the right places, they can add considerable ambiguity to your sentences. 'I saw three squid scuba-diving last Thursday' conjures up a very interesting image. Don't make readers guess what you're trying to say: they often guess incorrectly. Good scientific writing is precise, clear and unambiguous to the reader, whose mind may run on different lines from those of the writer.

7. **Never make the reader turn back.** Try to capture the reader's attention in your first paragraph and lead him or her through to the end, line by line, paragraph by paragraph. Avoid making the reader turn back two pages, or even one sentence. Link your sentences carefully, using such transitional words as 'Therefore' or 'In contrast', or repeating key words, so that a clear argument is developed logically. Remind the reader of what has come before, as in the following example.

> In saturated air (100% relative humidity), the worms lost about 20% of their initial body weight during the first 20 hours but were then able to prevent further dehydration. In contrast, worms maintained in air of 70-

> 80% relative humidity experienced a much faster and continuous rate of dehydration, losing 63% of their total body water content in 24 hours. As a consequence of this rapid dehydration, most worms died within the 24-hour period.

Note that the second and third sentences in this example begin with transitions ('In contrast, . . .', 'As a consequence of . . .'), thus continuing and developing the thought initiated in the preceding sentences. A far less satisfactory last sentence might read, 'Most of these animals died within the 24-hour period.' Link your paragraphs in the same way, using transitions to continue the progression of a thought, reminding the readers periodically of what they have already read.

Avoid casual use of the words *it, they* and *their.* For example, the sentence 'It can be altered by several environmental factors' forces the reader to go back to the preceding sentence, or perhaps even to the previous paragraph, to find out what *it* is. Changing the sentence to 'The rate of population growth can be altered by several environmental factors' solves the problem. Here is another example:

> Our results were based upon observations of short-term changes in behaviour. They showed that feeding rates did not vary with the size of the caterpillar.

The word *they* could refer to 'results', 'observations', or 'changes in behaviour.' The reader can go back to puzzle out what *they* are, but you should avoid the 'You know what I mean' syndrome. Changing *they* in the second sentence to 'These results' avoids the ambiguity and keeps the reader moving in the right direction. Do not be afraid to repeat a word used in a preceding sentence if it avoids ambiguity.

8. **Don't make readers work harder than they have to.** If there is interpreting to be done, you must do it. For example, never write something like: 'The difference in absorption rates is quite clearly shown in Table 1.' Such a statement puts the burden of effort on the reader.

Instead, write something like: 'Clearly, alcohol is more readily absorbed into the bloodstream from distilled, rather than brewed, beverages (Table 1).' The reader now knows exactly what you have in mind and can examine Table 1 to see if he or she agrees with you.

9. **Be concise.** Give all the necessary information, but avoid using more words than you need. By being concise, your writing will gain in clarity. Why say: 'Our results were based upon observations of short-term changes in behaviour. These results showed that feeding rates did not vary with the size of the caterpillar' when you can say the following? 'Our observations of short-term changes in behaviour indicate that feeding rates did not vary with the size of the caterpillar.'

   In fact, you might be even better off with the following sentence: 'Feeding rates did not appear to vary with the size of the caterpillar.' With this modified sentence, 50 per cent of the words in the first effort have been eliminated without any loss of content. The savings are not merely aesthetic. Authors are often asked to bear some of the printing costs of their papers, so it can literally pay to be concise. Cutting out extra words means you have less to type, and your readers can digest the paper more easily.

10. **Stick to the point.** Delete any irrelevant information, no matter how interesting it is to you. Keep it for later use if you wish, but don't let asides interrupt the flow of your writing.

11. **Write for your fellow students and for your future self.** It is difficult to write effectively unless you have a suitable audience in mind. It helps to write essays or papers that you can imagine being understood by your fellow students. You should also prepare your writings so that they will be meaningful to *you* should you read them far in the future, long after you have forgotten the details of the experiments. Addressing these two audiences – your fellow students and your future self – should help you to write clearly and convincingly. For a lab report, imagine that you may have to teach or demonstrate a similar experiment in future.

12. **Don't plagiarise.** Express your own thoughts in your own words, as you will have to in exams. If you are quoting from another writer or restating that writer's ideas or interpretations, you must credit your source explicitly. Using quotation marks indicates a direct quotation. Note, too, that simply changing a few words here and there, or changing the order of a few words in a sentence or paragraph, is still plagiarism. Plagiarism is one of the most serious crimes in academia. It can get you expelled or cost you a career later. With practice and conscientious effort, you will find yourself capable of generating your own good ideas and presenting them in prose of your own devising.

13. **Don't be teleological.** That is, don't attribute a sense of purpose to other living things, especially when discussing evolution. Giraffes did not evolve long necks 'in order to reach the leaves of tall trees.' Cyanobacteria do not contain intra-cellular gas vacuoles 'to enable cells to float.' Evolution proceeds through genetic changes, chance, differential survival and reproduction, not with intent. Long necks, gas vacuoles and other genetically-determined characteristics may well have given some organisms an advantage over others in surviving and reproducing, but this does not mean that these characteristics were deliberately evolved in order to achieve something.

    Organisms cannot evolve structures, physiological adaptations or behaviour out of desire. Appropriate genetic combinations must arise by chance before natural selection can operate. Don't write, 'Insects may have evolved flight in order to escape predators.' Instead, write, 'Flight in insects may have been selected for in response to predation pressure.' Don't write, 'The parent gulls remove the white, conspicuous eggshells in order to protect the newly hatched, black-headed young.' Instead, write 'Parental removal of the white, conspicuous eggshells may help to protect the newly hatched, black-headed young gulls from predation.'

14. **Always underline or italicise species names:** for example, Homo sapiens or *Homo sapiens.* If you can use italics on a printer, do so. Note also that the generic name (*Homo*) is capitalised whereas the specific name (*sapiens*) is not. Once you have given the full name of

the organism, the generic name can be abbreviated; *Homo sapiens*, for example, becomes *H. sapiens*. It is not permissible to refer to an organism using the generic name, since most genera include many species. Always check the spelling of unfamiliar names. In some contexts and some journals, the naming authority (or the abbreviation) is also given, *e.g. Viola odorata* L., where L. stands for Linnaeus, who named this violet. *Viola riviniana* Ssp. *minor* (Murbeck) Valentine is an example of a subspecies name with the authorities quoted. For different groups of organisms, follow the latest editions of such works as *The International Code of Botanical Nomenclature*, *The International Code of Nomenclature for Cultivated Plants*, and *The International Code of Zoological Nomenclature*. Your librarian should be able to help you locate a copy. As a general reference, there is *Biological Nomenclature: Recommendations on Terms, Units and Symbols*, The Institute of Biology, London, 1989.

15. **Proof-read.** Although it is an important part of the writing process, none of us likes proof-reading. By the time we have arrived at this point in the project, we have put in a lot of work. Who wants to read the paper yet another time? Moreover, finding an error means having to make a correction. But put yourself in the position of your lecturer, who may read a hundred or more essays each term. He or she starts off on your side, wanting to see you earn a good mark. Similarly, a reviewer or editor of scientific research manuscripts starts off by wanting to see the paper under consideration get published. A sloppy paper – for example, one with many typographical errors – can lose you a considerable amount of good will; it suggests to the reader that you take little pride in your efforts. Furthermore, it is insulting. Failure to proof-read and correct your paper implies that you don't value the reader's time, which is a most unwise message. Never forget: there is often a subjective element to grading and to decisions about the fate of manuscripts and grant proposals. In addition, a carelessly proof-read paper may suggest that the research was carelessly performed. For all these reasons, shoddily-prepared material can easily lower a grade or mark, damage a writer's credibility, reduce the likelihood that a manuscript will be accepted for publication or that a

grant proposal will be funded, or cost an applicant a job or a place for further studies. An effective way to proof-read is to read the work aloud, when clumsy sentences, errors of punctuation and grammar, and very dull passages may stand out vividly.

16. **Appearances are important.** Your writing should give the impression that you are well-organised, that you took the assignment seriously, that you are proud of the result, and that you welcome constructive criticism. See below for when to use handwriting, typing or word-processing, but whichever you use, the result must be neat, orderly, well set-out, and legible. Use only one side of each page; this is very useful if you have to do a scissors-and-sticky-tape-revision later. On each page, leave margins of about 1½ inches (4 cm) on the left side and about an inch (2.5 cm) on the right side and at the top and bottom. Double-space your typing so that your lecturer can easily add comments at any point. Make corrections neatly. Never underestimate the subjective element in grading. Use clear drawings, graphs or other illustrations where relevant. A clear, well-labelled sketch can often convey information much better than many words.

17. **Organise your work systematically,** for ease of finding things and for clarity. For example, put your name and the date at the top of each assignment. Pages should be numbered so that the reader can tell immediately if a page is missing or out of order and can easily point out problems on particular pages ('In the middle of page 7, you imply that . . .'). Keep the pages of an essay or lab report together by means of a labelled folder, binding, tag or other secure fastening. File your notes and references systematically for easy retrieval, in clearly-labelled folders. Numbering the pages of your lecture notes and writing those numbers down in the appropriate places on the course lecture synopsis can provide a handy index to material in a course, for easy retrieval of particular facts.

18. **Make a copy of your work before submitting the original.** Lecturers sometimes lose or mislay things, and you may need to refer to the work before the original is handed back.

19. **Buy a big dictionary and use it frequently.** You will need a dictionary to check spellings and word meanings. Buy a large one which also includes word origins, pronunciation, syllables and stresses: Chapter 5 will show you why all these items are useful for your writing. I (BL) have a large Chambers 20th Century Dictionary at work and one at home, with some other dictionaries for comparisons and cross-checking. I make frequent use of these dictionaries for biology and other interests.

20. **Learn to touch-type.** As early in your career as possible, learn to type with all ten fingers and without having to look at the keys. Either attend a course, or use the appropriate teaching program on a word-processor. It will save an enormous amount of time and effort while you are a student, and in your later career.

21. **Make friendly use of the staff.** We are there to help you. Like you, we are overworked and underpaid, but many of us actually enjoy helping responsive students. Always enquire if you are not sure about the expected length or format of requested written work, or how to get started. Staff have brains, knowledge and experience, all of which can be of use to you. That includes people like technical staff, whose help is often extremely valuable in projects or practicals, when you may have to borrow equipment or chemicals, or obtain lab space, or use unfamiliar techniques. Demonstrators, research students and students from previous years can also be very useful if approached correctly. A student who appears friendly, interested and willing to learn can be a joy to teach and to help; one who appears surly, truculent, bored and disruptive is unlikely to get such ready help, or good references. Make it clear that you realise that staff are busy all the time, and ask when it would suit them to give you the help you need. 'Is this a convenient time to discuss my essay?' is preferable to 'Are you busy?'

WHEN TO USE HANDWRITING, TYPING, AND COMPUTER WORD-PROCESSING AND GRAPHING

For ordinary tutorial essays and most lab books, legible handwriting is quite acceptable and is frequently used in most institutions of higher and further education. As students normally answer exam questions in handwriting, such essays and practical books are a good way of getting feedback on the legibility of your writing. Do take any criticisms of legibility very seriously, as illegibility in exams loses many marks. For undergraduate research projects, clear handwriting, typing and word-processing are usually all acceptable, with word-processing becoming the most common. For a formal MSc, MPhil or PhD thesis, typing or word-processing is normally obligatory. If in doubt as to what is acceptable, consult the appropriate member of staff.

Try this handwriting exercise, now, with your usual pen. Write the following words and numbers in your normal handwriting, at your normal speed:
urine, wire, wine; Ray, Roy; rag, ray, raj; Anthurium, Arthurian; diner, dinner, dimmer; dear, clear; ball, bull, lull, bell, boll, bill, brill; aluminium; core, cane, care, come, came; anemophilous; aneurin; chromatid, chromosome; dilute, dilate; voraciously, veraciously; liaison; frugiferous; pale, pate; pit, pet, put, pat, pot; force, farce; anther, antler; amylase, amylose; ZOO, 200; 461732589.

Now check your writing for the following points. Where there are groups of similar words, is each word legibly distinct from the others? If you cover up the other words within a set, can a friend read which you have written? Can you always distinguish *a* (low initial and final strokes, closed top) from *o* (high initial and final strokes, closed top) and from *u*? Can you and your friend always distinguish your *t/l, g/y/j, r/n/m/, i/e, d/cl, bi/bu/lu/br*? To confuse enzymes with their substrates is a major scientific error, so make sure that enzymes such as *amylase* are quite distinctly different from their substrates, such as *amylose.* Are your numerals all distinct? If your 1 and 7 are too similar, adopt the continental 7̶. Where you have symbols involving letters and numbers, as in maths and in some strain designations, using Ø for zero can help to distinguish that from capital letter O. See what needs to be changed in your letters and numerals to make them more distinct from each other.

When I (BL) am writing for a secretary and think that a word (technical or non-technical) might be unfamiliar, or might be misread from my not-very-good handwriting, I put that word in capitals above the hand-written word, for extra clarity. I also read through what the secretary has typed even if she is copying from typewritten material, as errors do happen, such as *a new kind* once being rendered as *a new king.*

USING COMPUTERS IN WRITING, ANALYSIS AND PRESENTATION

Computers used for word-processing are wonderful for revising drafts. Back in the days of typewriters, when reading drafts of my (JP) papers I would often see places where rearranging a few paragraphs, adding a phrase or sentence, or even replacing one word with another, would have substantially improved the final product. But if I had already typed several drafts, I rarely made those additional changes because of the unbearable thought of retyping pages yet again. With word-processing, however, perfection is within your grasp. It is now easy to change a word, modify or delete a sentence, or reorganise a paragraph or an entire paper, and the computer will produce the revised version at the touch of a button.

Even so, it is best to write your first drafts with pen and paper. First drafts serve primarily to get ideas on paper, where they can't escape; the form, order and manner of expression are not major concerns at this early stage. Consequently, the first revision is often so extensive that it is far less time-consuming to revise this draft by hand than to word-process one's way through it. In fact, putting a first draft on the computer might actually inhibit you from making the extensive revisions that are called for. Moreover, if there is a power failure or if your disk gets mangled in the midst of your brilliant insights, you can lose everything; if you use a pen instead, you will still have your first, handwritten draft. The second draft can be entered into the computer.

If you have never used a personal computer for word-processing, do not be intimidated by the jargon used by those who have. Getting started is easy, and there is usually a 'Tutorial' demonstration. There are only a few basic steps and there will probably be a lot of people around who can help you if necessary. Do regularly 'save' your work on the hard disk, in case of electrical problems or someone switching off the machine prematurely, and always back up your work on a floppy disk in case something goes wrong with the hard disk, or another user wipes out your file. In the case of a long document like a PhD thesis, keep more than one back-up copy on floppy disk, perhaps two at work and one at home in case one place burns down or gets flooded. Keep an accurate record of what is in each file and on each disk, as you must be sure not to replace a later version with an earlier one by accident. If you have some earlier versions on disk, you can re-insert material if you change your mind about what you previously deleted from your main copy. Hard disks do 'crash', and material on them may not be recoverable, as I (BL) discovered to my cost.

Learning to Edit With a Computer

Learning to edit with a computer is only confusing if you try to learn everything at once. At the start you need to learn only a few basic commands: move left within a line, move right within a line, move up a line, move down a line, delete a letter or space, add a new letter or space, and reformat (tidy up) a paragraph after adding or deleting material. After you gain experience, you can learn trickier manoeuvres, such as deleting entire words or sentences with a single punch of a button, or moving sentences and paragraphs from one place to another. For ordered displays such as tables, you will probably need Left Tab, Right Tab and Decimal Tab.

Be warned, though, about what you cannot expect a computer to do. Advertisers say that with your personal computer you will see your spelling improve, your sentences make sense, your paragraphs become well organised, your ideas seem brilliant and your grades soar! As readers of many computer-printed reports and papers, we must inform you that computers do not work these kinds of miracles. Word-processing programs cannot remove from the author the responsibility for thinking, organising, revising and proof-reading. In particular, they can do little to help you in the first struggle – the struggle for understanding – and they cannot think, organise or revise for you.

Computerised spelling checkers are of some use in catching typographical and spelling errors, but will not catch all mistakes. Biology is a field with much specialised terminology, most of which is of no use to non-biologists; its terms, therefore, do not often find their way into spell-checker dictionaries. Although you can easily add words to the computer's dictionary, the terminology you need may be changing with every new assignment. Moreover, a spelling-checker program will not distinguish between *to* and *too*, *there* and *their*, or *it's* and *its*, and will miss typographical errors that are real words; using the program will not spare you the chore of proof-reading for spelling mistakes. Suppose for example that you typed *an* when you intended to type *and*, or you typed *or* when you should have typed *of*, or *rat* instead of *rate*. To find these errors you would have to use a program specialising in catching grammatical mistakes, but such programs will not catch every error and do not always suggest the proper correction when mistakes are recognised. By all means use spelling- and grammar-checking programs if they are readily available, but then use your own sharp eyes and keen intellect to complete your proof-reading. Do not rely on machines to do for you what you have to do for yourself in an exam, without such a machine. If you use a shared computer or printer, do note that several people may desperately want to use it at the same time as you, to meet the same deadline, so make allowances for that in your planning.

Word-processing is a two-edged sword; because revisions no longer require time-consuming retyping, use of a computer places increased responsibility on the writer to see that the revisions get made. Lecturers find it annoying when students turn in computer-printed reports that are carelessly written and not proof-read.

In addition to their use as word-processors, computers are used by many biologists for data storage or analysis and for statistical evaluation of large or complex data sets. But biology undergraduates will often find that a scientific calculator will be perfectly adequate for most calculations.

Computers can be a real asset in preparing graphs and tables, as discussed in Chapter 3, pages 73 and 83. Improved 'packages' for graphics are now widely available, some of them easy to use, but make sure that you know how to use them properly, not leaving that to the last minute. Many students suffer each year through struggling with an unfamiliar graphics program to display their project data on the night before the report has to be handed in, often with unfortunate results. You should not feel compelled to generate your figures and tables by computer; hand-drawn graphs and tables should earn you as good a mark, provided they are carefully planned, sensible and neatly executed. If your course involves a lot of mathematical biology or statistics, the staff can advise you on what you need. With the advent of computing labs, work-stations and networking, the hardware and software needed, say for the mathematical modelling of some biological process, will usually be provided by your department, usually with the necessary training for use and for presentation of the results.

SUMMARY

1. Acknowledge the struggle for understanding and work to emerge victorious; read with a critical, questioning eye.
2. Think about where you are going before you begin to write, while you write and while you revise.
3. Write to illuminate, not to impress.
4. Back up statements of fact or opinion where relevant, especially in theses.
5. Always distinguish fact from possibility.
6. Write exactly what you mean.
7. Never force the reader to turn back.
8. Don't make readers work harder than they need to.
9. Be concise: avoid unnecessary words and stick to the point.
10. Write for an appropriate audience: your fellow students and your future self.
11. Don't plagiarise.
12. Avoid teleology.
13. Underline or italicise the scientific names of species.
14. Proof-read all work before handing it in, and keep a copy for yourself.
15. Make your papers neat in appearance, double-space all work, and leave margins for the lecturer's comments and suggestions.
16. Organise your work systematically. Put your name and the date at the

top of each assignment, number all pages, and label and index your files, disks and folders.

17. When possible, use word-processing computer programs for revising drafts of manuscripts, major project reports and theses, but prepare the first draft or two with pen and paper. Clear handwriting is acceptable for many essays and reports – if there is doubt, seek guidance from staff.
18. Buy a large dictionary and use it frequently.
19. Learn to touch-type as early as possible in your career. It will save you a lot of time and trouble.
20. Make friendly and considerate use of a whole range of staff; they were students once!

2. General Advice on Reading, Lectures and Note-Taking

EFFECTIVE READING

Too many students think of reading as the mechanical act of moving the eyes left to right, line by line, to the end of a page, and repeating the process to the end. When the last page has been 'read', the task, these students believe, is over and they go on to something else. This is 'brain-off' reading. In the same way, many students use 'brain-off' writing, 'listening' to a lecture by furiously copying whatever the lecturer writes or says, without really thinking about the information presented. However, if you hope to develop something worth saying in your writing, you must *interact* intellectually with the material; you must become a 'brain-on' reader and writer, thinking actively about every sentence, every illustration, and every table. If you don't fully understand any part of what you are reading (including your lecture notes), you must work on the problem until it is resolved.

This is time-consuming but there are things you can do to smooth the way. Whether you are writing an essay or a laboratory report, begin by reading the appropriate sections of your textbook and lecture notes to get a solid overview of the general subject. It is usually wise to consult one or two additional textbooks. If you need to venture into the primary literature, a firm foundation is required. Your lecturers may have placed a number of pertinent textbooks on reserve in your library, or have provided a reading list. Alternatively, you can consult the library filing system, looking for books listed under the topic you are investigating. There is more about locating appropriate sources later in this chapter.

Primary literature consists of original research results in scientific journals, research reports (*e.g.* annual reports of institutions such as the AFRC Institute of Grasslands and Animal Production), conference proceedings, theses and patents. Secondary sources are textbooks, review articles, data compilations, abstracts, indexes and bibliographies (descriptive lists of books in a particular subject area). Tertiary sources include guides to abstracts and indexes, bibliographies of bibliographies, and catalogues, such as your library catalogue, publishers' lists, lists of books in print, etc. First-year students mainly use books and reviews. In later years you make more use of primary sources such as original research papers, and may need to do literature searches, as described later in this chapter. Post-graduates can still find useful information in recent editions of specialist textbooks, but will make extensive use of original papers, and perhaps theses of previous post-graduates in their department, or elsewhere.

TEXTBOOKS

You may be given a general list of textbooks for a year or for a specific course. If the list does not make it clear, ask the staff which books should be bought and which are only for occasional reference. You may not be able to afford all the recommended books, so try to establish which are the key ones and whether it is necessary to buy the latest edition. There may well be students from previous years wishing to sell their copies to you: check any relevant noticeboards or advertisements. Sometimes it is important to have the latest edition, as science moves so rapidly. Make sure that you register to use your library and that you know how many books you can borrow and for how long. When considering whether to buy or borrow a textbook, remember that everyone on a course is likely to want the library copies at the same time, and that library copies may be on a reference-only or restricted-loan basis when they are most needed. With your own copy, you can even highlight important passages and make notes in the book, although these could affect the resale value later.

Read your textbooks for general background to your course/s and to check specific points from lectures, as well as for information for essays. Look through the table of contents at the front for general areas of interest and the index at the back for specific topics. If there is a glossary defining the terms used, this can be very helpful when you meet unknown terms, some of which will not be in a general English dictionary. Do read the text thoroughly; don't just look at the illustrations. Many textbooks have problems or questions at the ends of chapters, allowing you to check whether you have understood and can use that chapter's material. These are often worth trying unless your course includes similar tests. Some books have answers at the back, while others have separate answer manuals.

TAKING NOTES DURING LECTURES

Be there on time, with plenty of paper, spare pens, pencils, coloured pens and a ruler. Sit where you can see the board or boards easily. Get yourself in the frame of mind to concentrate on the lecture: do not eat or drink during it, or do a crossword, read a newspaper, chatter with your friends, or suffer from lack of sleep or from a hangover. Remember that it is costing someone a lot of money to pay for your studies, so take full advantage of the teaching offered. Have the lecture synopsis in front of you so that you can see where each part fits into the general plan of the course. If you have been given handouts on the course, try to read the relevant parts before the lecture.

During the lecture, concentrate hard, trying both to understand what is being said and to take down the essential facts. There is no need to write in full sentences; notes are sufficient. If there is a bit you cannot get down quickly enough, leave a gap in your notes and try to fill it later by consulting other students and your textbook. If you are a slow writer, try to speed up with practice, and make friends with a fast writer, whose notes you can later use to fill in the gaps. It takes practice to understand and to write simultaneously, but it is a skill you will often need in future. Writing is such a help in concentrating one's thoughts that we, the authors, sometimes take notes on a talk even though we are unlikely to need those notes later; the act of note-taking forces us to concentrate on the essentials. If you have been given extensive handouts for the lecture, it is best to concentrate on what the lecturer is saying, rather than trying to read the handout at the same time. There is no point in copying down diagrams, tables or large factual chunks if they are already given in the handout, which you can cross-reference to your lecture notes. It is difficult to take adequate notes if the lecturer is showing lots of slides in a darkened lecture theatre, or presenting a long series of detailed overhead transparencies very quickly. It is quite acceptable to ask the lecturer how much of such material you are expected to know, and how much is just illustrative background.

Do make full use of different colours when taking down diagrams. For example, a different colour for each of the four chromatids in diagrams of multiple crossovers at meiosis can make the final result much easier to follow than if you use only one or two colours. If you take down graphs, remember to label both axes.

We advise our students to sort out any difficulties with lectures as soon as possible, from the textbooks or by asking us, because unresolved problems can affect the understanding of later lectures or get forgotten. It is often convenient for students to ask their questions at the end of the lecture, or during the next practical class, or during tutorials. Asking questions during a lecture tends to disrupt the flow and perhaps cause late-running, but asking during the lecture is fine if a word is illegible on the board or inaudible, or is unknown to the class. Lecturers do make slips while lecturing, and reasonable lecturers are happy to have such slips pointed out at the time, especially if this is done in a tactful manner. Many problems can easily be resolved from the appropriate textbook. You should aim to have your lecture notes in their final form within a week of a lecture being given, with all queries resolved, any illegibilities or omissions corrected, all unfamiliar words defined, and with the notes in a suitable form for learning and revising from. Think from the start about your need for revision, and don't leave its organisation until just before the exams! If you have a handicap such as dyslexia, make sure that staff are notified, perhaps after showing a medical certificate to the senior tutor or course convenor; in such cases, you may be allowed extra time in exams. For people with special difficulties, lecturers may be willing to lend them a photocopy of their lecture notes, which those students can then photocopy for themselves.

READING SCIENTIFIC PAPERS

Primary scientific literature must be read slowly, thoughtfully and patiently, and a single paper must usually be reread several times before it can be thoroughly understood; don't become discouraged after only one or two readings. Reading scientific literature is slow going but gets easier with practice. If, after several rereadings of a paper and after consulting several textbooks, you are still baffled by something, ask your lecturers for help.

As you carefully read each paper, pay special attention to the following:

1. What specific question is being asked?
2. How does the design of the study address the question posed?
3. What are the controls for each experiment?
4. How convincing are the results? Are any of the results surprising?
5. What contribution does this study make towards answering the original question?
6. What aspects of the original question remain unanswered?

Data – the most important part of any book or journal article – are displayed as figures, photos, or tables; it is important to develop the skills needed to examine these elements critically. Your goal is to reach your own interpretation of the data so that you can better understand or evaluate the author's interpretation. To do this, you must study the data and ask yourself questions about how the study was done, why it was done and what the major findings were.

Consider the example shown in Figure 1 (next page), modified from a 1990 review entitled 'Peptide Regulation of Mast-Cell Function', by D. E. Cochrane. From your background reading or lecture notes, you would probably know that mast cells release into the blood a variety of substances involved in provoking allergic and inflammatory responses. If you didn't already know this, you would do some background reading of your textbook.

Looking at Figure 1, let us see if we can deduce what the researchers did to collect their data. By reading the axis labels, we learn that the graph shows how blood histamine concentrations change over time, and

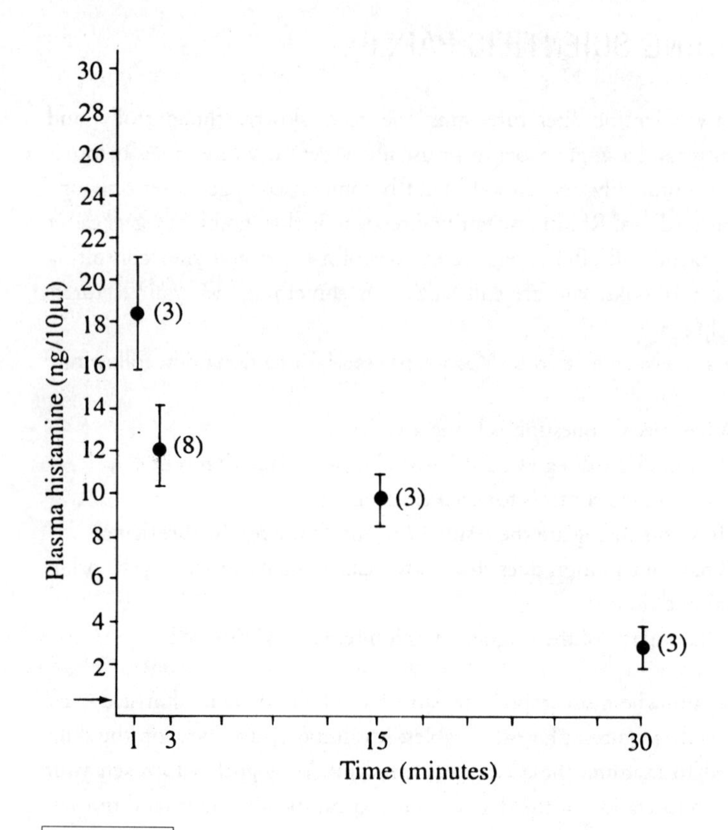

FIGURE 1.

Plasma histamine levels in response to neurotensin (NT) given at t=0. Rats were anesthetized and given NT (5 nmol/kg) or saline (0.3 ml) intravenously. Blood samples were collected at the indicated times. Each point represents the mean (± one standard error about the mean) of n values (given in parentheses). The horizontal arrow (lower left) shows the mean histamine concentration before addition of NT. Intravenous injection of saline did not alter this concentration over the 30-minute observation period. From D. E. Cochrane, 1990. Peptide regulation of mast-cell function. In, *Progr. Medicinal Chemistry*, Vol. 27; G. P. Ellis and G. B. West, eds.; Elsevier Science Publ. (Biomedical Division), pp. 143-188.

we learn from the figure caption that these changes are provoked by a particular peptide called neurotensin (NT). The study was done on anaesthetised rats and the action seems to have occurred quickly, as the X-axis extends only to 30 minutes. Looking more closely, we see that histamine levels were initially quite low (less than 1 nanogram per 10 microlitres of blood plasma), as indicated by the arrow at the lower left side of the graph, and that they rose impressively by the time the first blood sample was taken, one minute into the study. Other blood samples were taken 3, 15, and 30 minutes after neurotensin was administered and three to eight separate samples were taken at each time period. Even without reading the figure caption and without seeing the numbers alongside each data point, we would know that replicate samples were taken, since the thin lines extending vertically from each point indicate the amount of variation about the mean value for each sampling time; one can see variation about a mean value only when multiple samples are taken.

What were the controls for this experiment? A number of rats were injected with a saline solution instead of with neurotensin, and blood samples were also taken from these rats at appropriate times. So, we know quite a bit about how this aspect of the study was conducted, just from scrutinising the graph. It helps considerably that the graph and figure caption were carefully constructed. What additional information might you wish to have? Here are a few questions that might arise as you think about the figure.

1. How did Dr Cochrane decide to inject NT at a concentration of 5 nmol/kg? Is this a physiologically-realistic concentration?
2. How was histamine concentration determined?
3. How many rats served as controls for each time interval?
4. How old were the rats?
5. What sex and strain were they?
6. How were the rats anaesthetised, and might this pretreatment affect the response to NT?
7. With regard to the replicate samples, were three to eight blood samples taken simultaneously from a single rat at each sampling time, or were blood samples taken from three to eight different rats each time?

Blood samples were *probably* taken from a number of different individual rats, but we can't be sure of that from the graph.

8. Were different rats sampled at each time interval, or were the same individuals sampled repeatedly, for example, at 15 minutes and again at 30 minutes? One might guess that the same individuals were sampled repeatedly. On the other hand, there may be technical limitations to drawing blood from any individual rat more than once within a certain timespan. Suppose, for instance, that a very large volume of blood was required to determine the histamine concentration. This issue leads to the last two questions.
9. What volume of blood was withdrawn for each sample?
10. How were the blood samples obtained?

Asking such questions puts you in a good position to interpret the results illustrated and provides a framework for your later reading of the Materials and Methods section of the report; you can now enter the methodology section of the paper looking for the answers to specific questions. If asked to describe how this aspect of the study was conducted and why the study was probably undertaken, how might you respond? Even without reading anything else in the paper, you could write the following summary.

> This study was apparently undertaken to determine the ability of the peptide neurotensin to elicit histamine secretion from mast cells. A number of rats were anaesthetised and then injected intravenously with 5 nmol/kg of neurotensin, while others (controls) received intravenous injections of saline solution to control for the possibility that the act of injection itself provoked histamine release. Some quantity of blood was withdrawn from a number of rats 1, 3, 15, or 30 minutes after injections were made, and the histamine concentration in each blood sample was somehow determined.

Now, in a few short sentences, let us try to summarise the results, beginning with the most general statements we can make.

1. Neurotensin had a dramatic and rapid effect on histamine concentrations in the blood of laboratory rats, with histamine concentration increasing about 20-fold within one minute after injection.
2. The effect seems rather short-lived; only two minutes into the study the histamine concentrations had already fallen considerably from the peak level recorded one minute earlier.
3. By 30 minutes after the injection, histamine concentration approached initial levels.
4. The effect shown is clearly due to the peptide rather than the procedure itself, because histamine concentrations in the blood of control rats receiving saline injections did not change appreciably during the study.

You can approach tabular data in the same way, first reading the accompanying legends and column headings, and then asking yourself *how* the study was done, *why* it might have been undertaken and what the key general results seem to be. Apply the same procedure graph by graph, table by table, until you have finished the last figure. Now it is safe actually to read the text of the paper! Perhaps you will find the author or authors reaching conclusions different from your own. Perhaps you missed some crucial element in studying the data, or you will learn something crucial in the text that was not clear in the figure. In this case, you can happily leave your own opinion behind and embrace the author's. On the other hand, you may genuinely disagree with the conclusion reached by the paper's author, or you may object to the author's interpretation because of some concern you have about the methodology used. This is, in fact, how new questions often get asked in science, and how new studies get designed, each new step building on the work of others. Most of us will never have the final say in any particular research area, but we can each contribute a valuable next step, even if our individual interpretation of that step later turns out to be mildly or even dramatically wrong.

You will now be in an excellent position to *discuss* the paper, on its own or in relationship to other studies that you will go on to scrutinise as thoroughly. If you don't go through these steps in 'reading' the data, you will be all too accepting of the author's interpretation. In consequence, you will have difficulty avoiding the book report format in your writing, simply repeating what someone did and what they say they found. You can move your work to a higher, more interesting plane (more interesting for you and the reader) by becoming a 'brain-on' reader. It's tough slogging and although it becomes easier with practice, it never becomes trivial. But it does become fun, and satisfying, in the way that playing a good tennis game in hot weather can be fun and satisfying.

PLAGIARISM AND NOTE-TAKING

The work you submit for evaluation must be original: yours. Submitting anyone else's work under your name is plagiarism and can get you expelled. Presenting someone else's ideas as your own is also plagiarism. Consider the following two paragraphs.

> Smith (1981) suggests that this discrepancy in feeding rates may reflect differences in light levels used in the two different experiments. Jones (1984), however, found that light level did not influence the feeding rates of these animals and suggested that the rate differences reflect differences in the density at which the animals were held during the two experiments.

> This discrepancy in feeding rates may reflect differences in light levels. Jones (1984), however, found that light level did not influence feeding rates. Perhaps the difference in rates reflects differences in the density at which the animals were held during the two experiments.

The example of the first paragraph is fine. In the second example, however, the writer takes credit for the ideas of Smith and Jones; the writer has plagiarised.

Plagiarism sometimes occurs unintentionally, through faulty note-taking. Photocopying an article or book chapter does not constitute note-taking; neither does copying a passage by hand, occasionally substituting a synonym for a word used by the source's author. Take notes using your own words; you must get away from being awed by other people's words and move towards building confidence in your own thoughts and phrasings. Note-taking involves critical evaluation; as you read, you must decide either that particular facts or ideas are relevant to your topic or that they are irrelevant. If an idea is relevant, you should jot down a summary using your own words. Try not to write complete sentences as you take notes; this will help you avoid unintentional plagiarism later and will encourage you to see through to the essence of a statement while note-taking.

Sometimes the authors' words seem so perfect that you cannot see how they might be revised to best advantage for your paper. In this case, you may wish to copy a phrase or a sentence or two verbatim, but be sure to enclose this material in quotation marks as you write, and clearly indicate the source and page number from which the quotation derives. If you modify the original wording slightly as you take notes, you should indicate this as well, perhaps by using modified quotation marks: " . . . ". If your notes on a particular passage are in your own words, you should also indicate this as you write. You might, for example, precede such notes, reflecting your own ideas or your own choice of words, with the word *Me* and a colon; or you could use your initials. If you take notes in this manner you will avoid the unintentional plagiarism that occurs when you later forget who is actually responsible for the wording of your notes, or for the origin of an idea.

Here is an example of notes taken using this notation. They are based on a paper published in 1989 by M. D. Bertness. Figure 2 (next page) shows an excerpt from the paper and Figure 3 shows some notes based on that excerpt. As shown in Figure 3 (page 31), the note-taker has clearly distinguished between his or her thoughts and the author's thoughts, and between what the author has done and what the student thinks could be done later or might have influenced the results (*e.g.* 'Me: Why is it impt. to do study in protected bay?'). Note that the student has avoided using complete sentences, focusing instead on getting the basic points and pinning down a few words and phrases that might be useful later. The student will not have to worry about accidental plagiarism

Introduction

The roles that intra- and interspecific competition play in shaping natural populations and communities has long been of interest to ecologists (Hairston et al. 1960, Strong et al. 1984). High population densities often result in competitive processes dictating natural distribution and abundance patterns (e.g., Buss 1986), but consumer pressure (Paine 1966, Harper 1969) and physical disturbance (Dayton 1971, Platt 1975) often ① reduce densities and minimize the importance of competition in nature. Limited recruitment (Underwood and Denley 1984, Roughgarden et al. 1987) and harsh physical conditions (Connell 1961*a*, Fowler 1986) also potentially limit population size, and reduce the impact of competitive phenomena on populations. ② High population densities, however, can also facilitate survival by buffering individuals from interspecific competitive pressures (Buss 1981), consumers (Atsatt and O'Dowd 1976), physical disturbance (Bertness and Grosholz 1985), and physiological stress (Hay 1981). Largely due to these contrasting effects of recruitment variation in natural populations and the variety of other factors that can independently or interactively influence populations, general statements relating recruitment density to population processes have been hard to make even in extensively studied systems (Connell 1985).

③ The study of plant and animal assemblages in hard substratum marine habitats has been valuable in understanding the interplay of biotic and physical factors in generating pattern in natural communities (e.g., Connell 1961*b*, Paine 1966, Dayton 1971). Sessile organisms in these systems are often space-limited, found ④ along gradients of environmental harshness, and easily trackable and amenable to experimental manipulation. Acorn barnacles are conspicuous members of many temperate zone intertidal communities, dominating a distinct zone at intermediate to high tidal heights (Stephenson and Stephenson 1949). Within this habitat, barnacles compete intra- and interspecifically by crushing and/or overgrowing neighbors (Connell 1961*a*, Wethey 1983*a*), are limited from higher intertidal habitats by heat and desiccation (Connell 1961*a*, Wethey 1983*b*), and from lower intertidal habitats by predators and competitors (Connell 1972, Menge 1976).

⑤ In this paper, I examine the intraspecific density-dependent dynamics of the barnacle *Semibalanus balanoides* in a wave-sheltered southern New England bay. I document variation in recruitment and survivorship, investigate the role of density-dependent mortality in generating survivorship patterns across the intertidal habitat, and examine physical constraints on barnacle success.

Methods

⑥ The *S. balanoides* population at the study site was examined in 10 × 10 cm quadrats on boulder (>1 m diameter) surfaces. A third of these quadrats were placed at low (−0.1 to 0.0 m, relative to mean low water) tidal heights, a third at intermediate (0.0 to +0.5 m) tidal heights, and a third at high (+0.5 to 1.0 m) tidal heights. Quadrats were individually numbered with aluminum tags attached to 6-mm (quarter-inch) stainless steel corner screws. Quadrats were established in February 1985 and followed until September 1987. All quadrats were ⑦ photographed monthly from March–September through 1986. In 1987 photographs were taken in April, after settlement, and in September. The only major barnacle ⑧ predator at the site, *Urosalpinx cinerea*, was removed from boulders containing quadrats in 1985 and 1986.

Excerpt from M. D. Bertness. 1989. Intraspecific competition and facilitation in a northern acorn barnacle population. *Ecology* 70: 257-268. Numbers in the margin correspond to the notes shown in the margin of Figure 3.

1) Low no. individs/m²[1] --→ reduced competition for resources.
Factors acting to keep pop. density low: physical stress, predation, low recruitment.
ME: What is "recruitment"? How defined?
2) Under some circumstances, survival is better when pop. densities are high; e.g., crowding can protect from physical stress. ME: Really? How?
3) Key issue: How do physical and biological factors interact to determine plant and animal distributions?
4) Acorn barnacles good to study since don't move around, are abundant intertidally, and are subject to various degrees of heat stress, dehydration stress, predation, and competition (including with other barnacles of same or other species)
ME: How many barnacle species are there? How many on one beach?
5) This study deals with one species (<u>Semibalanus</u> <u>balanoides</u>) from New England coast (protected area, no waves).
ME: Why is it impt. to do study in protected bay? Would expect diff. results in more wave swept area? Why? Or just easier to work w/o[2] having to deal with waves?
Looks at recruitment (ME: appearance of young barnacles on rocks?) [yes: see definit. at bottom left p. 258] and survival at diff. tidal heights. Hopes to explain findings with respect to temp. stress.
6) Methods: Sets up 10 × 10 cm quadrats on boulders at 3 diff. tidal heights. ME: if in highest tide area, barnacles exposed longer to air.

7) Follows barnacles in quadrats for about 2.5 yrs, by monthly photos.
8) Removes the major predator (<u>Urosalpinx</u> <u>cinerea</u>) monthly in first two years. ME: <u>U. cinerea</u> = "oyster drill" snail.

FIGURE 3.

Handwritten notes based on article by M. D. Bertness (see Figure 2). Numbers in margin correspond to the indicated portions of Figure 2.

[1] m² = metres squared

[2] w/o = without

when writing a paper based on these notes. Moreover, the student is well on the way to preparing a solid essay, since the style of note-taking indicates clearly that the student has been thinking while reading.

Some people take notes on index cards, with one idea per card so

It may be worth while to give another and more complex illustration of the action of natural selection. Certain plants ← (1) excrete sweet juice, apparently for the sake of eliminating something injurious from the sap: this is effected, for instance, by glands at the base of the stipules in some Leguminosæ, and at the backs of the leaves of the common laurel. This juice, though small in quantity, is greedily sought by insects; but their visits do not in any way benefit the plant. Now, let us ← (2) suppose that the juice or nectar was excreted from the inside of the flowers of a certain number of plants of any species. Insects in seeking the nectar would get dusted with pollen, and would often transport it from one flower to another. The flowers of two distinct individuals of the same species would thus get crossed; and the act of crossing, as can be fully proved, gives rise to vigorous seedlings which consequently would have the best chance of flourishing and surviving. The ← (3) plants which produced flowers with the largest glands or nectaries, excreting most nectar, would oftenest be visited by insects, and would oftenest be crossed; and so in the long-run would gain the upper hand and form a local variety. The flowers, also, which had their stamens and pistils placed, in relation to the size and habits of the particular insects which visited them, so as to favour in any degree the transportal of ← (4) the pollen, would likewise be favoured. We might have taken the case of insects visiting flowers for the sake of collecting pollen instead of nectar; and as pollen is formed for the sole purpose of fertilisation, its destruction appears to be a simple loss to the plant; yet if a little pollen were carried, at first occasionally and then habitually, by the pollen-devouring insects from flower to flower, and a cross thus effected, although nine-tenths of the pollen were destroyed it might still be a great gain to the plant to be thus robbed; and the individuals which produced more and more pollen, and had larger anthers, would be selected.

FIGURE 4.

Taken from Charles Darwin's *The Origin of Species,* published in 1859, from which notes on the mechanism of natural selection were taken (see Figure 5).

Plant evolution tied to insect behavior. Flowers now = effective in attracting insects for pollen exchange; how originate?

① Some plant sap apparently toxic (me: no evidence given). Plant nectar makes sap less nasty. Insects attracted to the sweet nectar, even though produced by plant originally to protect plant.

② If plant produces nectar in flower, insects attracted to flower, thus transport pollen, facilitate cross-fert. me: note that selection for nectar prod. in flower can occur only if a few plants accidentally start secreting nectar in flowers. Note that nectar not orig. prod. to attract insects; selected for protect plant, but once being produced, can evolve different function.

③ Those flowers that have greatest success attracting insects will spread the most pollen. me: now know that the genes of these flowers would thus incr. prob. of successful represen-tation in next generation.

④ Nectar prod. not essential to explain evol. of insect-mediated cross pollination: Suppose insect feeds on pollen (as some spp. do). Some pollen would stick to legs and be transferred to another flower. Again, flowers most successful in attracting insects would incr. chance of spreading genes, even though most of the pollen gets eaten.

FIGURE 5.

Handwritten notes based on the passage shown in Figure 4.

that the notes can be sorted readily into categories at a later stage of the paper's development. If you prefer to take notes on full-size paper, begin a separate page for each new source and write on only one side of each page to facilitate sorting later.

As you take notes, be sure to make a complete record of each source used: author(s), year of publication, volume and page numbers (if the source is a scientific journal), title of article or book, publisher, and total number of pages (if the source is a book). It is not always easy to relocate a source once it is returned to the library. Also, before you finish with a source, it is good practice to read it through one last time to be sure that your notes accurately reflect its content.

As another example of effective note-taking, consider some notes based on this paragraph (Figure 4, page 32) from Charles Darwin's *The Origin of Species*, which was published in 1859. The notes (Figure 5, page 33) were taken for an essay on the mechanism of natural selection. Notice that the student has taken notes selectively, that the notes are generally not taken in complete sentences, and that the student has found it unnecessary to quote any of the material directly and has clearly distinguished his or her own thoughts from those of Darwin.

It is hard to take notes in your own words if you do not understand what you are reading. Similarly, it is difficult to be selective in your note-taking until you have achieved a general understanding of the material. We suggest that you first consult at least one general reference text and read the material carefully, as recommended earlier. Once you have located a particularly promising scientific article, read the entire paper at least once without taking any notes. Resist the strong temptation to annotate and take notes during this first reading, even though you may feel that without a pen in your hand you are accomplishing nothing. Put your pen away and read slowly. Read to understand. Study the illustrations, figure captions, tables and graphs carefully; try to develop your own interpretations before reading those of the author(s). Don't be frustrated by not understanding the paper at the first reading; understanding scientific literature takes time and patience – and often many rereadings, even for experienced biologists. Concentrate not only on the results but also on the reason the study was undertaken and the way the data were obtained. The results of a study are real; their interpretation is always open to question. The interpretation is largely influenced by the way the study was conducted. Read with a critical, questioning eye. Many of the interpretations and conclusions in today's journals will be modified in the future. It is difficult to have the last word in biology; progress is made by continually building on and modifying the work of others.

By the time you have completed your first reading of the paper, you may find that the article is not really relevant to your topic. If so, the preliminary read-through will have saved you from wasted note-taking. Many students, in a hurry, photocopy articles, then use a highlighter pen to mark the most relevant passages after one rapid read-through. While better than not reading the paper at all, such a method is not fully 'brain-on' and gives much less understanding and insight compared with making notes as described above. Photocopying is useful for complicated diagrams or photographs, if you need them in your notes.

LOCATING USEFUL SOURCES

Once you have carefully read the appropriate sections of your textbook and the relevant portions of your lecture notes to get a solid foundation, you are ready to delve into more specialised material. The next step is to consult your library catalogue which will probably be a computerised one, with perhaps a section still on cards. If it is on microfiche (photographically much reduced in size onto transparent postcard-sized 'fiches'), the library will provide microfiche readers. The easiest subject approach is by keywords and it is advisable to try as many variations as possible. Suppose, for example, that you wish to find material on reptilian respiratory mechanisms. You might try, to no avail, Respiration or Reptiles, but looking under Physiology or Comparative Physiology might succeed.

USING A COMPUTERISED LIBRARY CATALOGUE TO LOCATE BOOKS

Most university libraries have computerised book catalogues, with terminals at various places in the library for student use. Here is an example of a search at the Imperial College Central Libraries, using a system called *Libertas* 6.0. The initial screen offers eight choices: 1, Quick AUTHOR/TITLE enquiries; 2, Full AUTHOR/TITLE enquiries; 3, NAME/AUTHOR enquiries; 4, TITLE enquiries; 5, KEYWORD enquiries (titles, subjects); 6, A to Z list of authors; 7, A to Z list of journal titles; 8, CLASSMARK enquiries. Suppose I am looking for books on chromosome structure. I press 5, for KEYWORD enquiries, followed by the 'enter' (return) key.

The new display instructs me to enter brief descriptions, so I enter 'chromosome structure'. The computer quickly informs me that there are 55 entries for 'Chromosome', 1000+ entries for 'Structure', 55 records found altogether for my joint requirements, and 4 items matching my search closely. It offers three further choices, of which I select 'display records'.

The next display shows details of five relevant books, with a choice of options, including displaying further lists, or locating particular books. The book on line 5 looks of interest: *The bacterial chromosome*, edited by Karl Drlica, Monica Riley, 1990, so I press LO, for locate, then 5 for item 5, as the computer asks which of the displayed book titles I am interested in. The next display tells me that it is available for 3-week loan from level 4, Life Science Library, at classmark 579.252 BAC, that the library has one copy and no copies are out on loan. I therefore know where to go to see or to borrow this book, and that there may well be related books nearby on the shelves around that classmark.

If I wanted to search the catalogue for other books around that classmark number, without leaving the terminal, I could return to the initial display, select option 8 for CLASSMARK enquiries, and find what book titles were at what classmark, perhaps asking for ones around 579. Using the catalogue has an advantage over looking on the library shelves as books out on loan and books on order are shown in the catalogue, and can be reserved from the terminal. Incidentally, some over-sized books may be placed on special shelves in the library, out of the normal sequence. One can also use the *Libertas* system to reserve a book, from the same terminal, and anyone connected to the campus computer network can do all this from his or her own computer.

REVIEW ARTICLES AND THE PRIMARY LITERATURE

The references given in textbooks often provide a good first access to primary literature, as do those given in review articles. Many review articles appear in the annual volumes of series such as *Annual Reviews of Entomology, Advances in Cell Biology, Biotechnology and Genetic Engineering Reviews,* etc. Besides reviewing the recent work in a particular subject area, each article also has a substantial bibliography which will give you many references to follow up. Many review articles also appear in periodicals such as *Biological Reviews, Bioscience,* and *Quarterly Review of Biology.*

Among the most useful and up-to-date are the 'Trends' and 'Current Opinions' series, *e.g. Trends in Genetics, Trends in Biochemical Sciences,* and *Current Opinion in Cell Biology.* These have a lively style and include short articles with many illustrations and diagrams. They concentrate on very current material and are ideal for learning about the latest research in the field covered. The series 'Critical Reviews in. . .' and 'Progress in. . .' are also useful.

It is profitable to browse through recent issues of journals relevant to your topic; ask your lecturers to name a few journals worth looking at. If you find an appropriate article in the recent literature, consult the literature citations at the end of the article for additional references. This is an easy and efficient way to accumulate references; the yield is usually high for the amount of time invested, although they will always refer to work which predates the article you are reading.

SEARCHING COMPUTER DATABASES

To expand your list of references and bring it up to date, it is necessary to search the abstracting and indexing journals. This aspect of literature searching has been completely changed by the recent electronic revolution in publishing. Not only are most of the major abstracts and indexes now available in electronic form, but recent developments in networking and CD-ROM (Compact Disk, Read-Only Memory) mean that this form of searching is no longer beyond the reach of undergraduates. In general, an academic library would be expected to hold some abstracts on CD-ROM for use at workstations in the library or, increasingly, over the campus network. Of interest to biologists are such titles as BIOSIS (the electronic form of Biological Abstracts) and MEDLINE (the electronic form of Index Medicus), although many databases in other fields contain relevant material and are worth investigating if time permits. Examples are MATHSCI, CITIS (Civil Engineering) and INSPEC (Physics, Electronics and Computing). ENVIROLINE covers environmentally-related publications; *Index to Scientific Reviews* indexes over 30,000 new reviews a year, and *Index to Scientific Book Contents* indexes individual chapters in multi-authored scientific books.

In the UK, many university libraries subscribe to BIDS (Bath Information and Data Service) which has mounted *Science Citation Index* and *Current Contents*, among other databases, so that they are accessible nationally over JANET, the Joint Academic Network. This means that the databases are available from any terminal connected to the campus network of participating institutions. The service is funded by subscriptions, usually paid by the library, and is normally free at the point of use.

Electronic searching relies on keywords for the subject approach but also allows other entry points such as title, institution, and in the citation indexes, authors. For earlier literature, predating the CD-ROMs and BIDS, there are many online services which access databases on commercial computers. These are, however, costly and more difficult to use, so are less suitable for undergraduates.

SEARCHING PRINTED VERSIONS OF ABSTRACTS AND INDEXES

Even though there has been such a movement to electronic forms, it is still useful to understand how to use the printed versions. The electronic databases are often constructed on the same principles, and the printed versions go back a lot further. Printed abstracts and indexes are often the only way to search early literature. Some notes on individual publications follow.

Using *Science Citation Index*

This index is unique in that it permits you to work forwards in time, finding relevant papers published later than the ones you already have. The *Science Citation Index* is published by the Institute for Scientific Information (ISI), Incorporated, in Philadelphia. Because it is expensive, not all libraries subscribe to this service.

To use *Science Citation Index*, you must first have discovered at least one paper (the so-called key paper) either from the primary literature pertinent to your quest, or from using keywords. Consulting the Citation volumes of the index, you look under the name of the author who wrote this key paper. Below that author's name you will find a listing of references, one of which should be for the paper you have already read and found to be particularly appropriate to your topic. Beneath the listing for this reference you will find a list of all the papers that have cited your key paper during the year covered by the index volume consulted. Suppose, for example, you have obtained and read the following reference, cited at the end of a chapter in your class textbook, and have determined the paper to be of special use in developing your topic: Kandel. E. R. and J. H. Schwartz. 1982. Molecular biology of learning: Modulation of neurotransmitter release. *Science* 218: 433–443.

Looking in the Citation volume for 1991 you will find the listing shown in Figure 6, which includes nearly 20 of E. R. Kandel's papers, published between 1961 and 1986, each of which has been cited at least once in the most recent year or so. The 1982 paper of interest to us is indicated by an arrow to the left of the reference. The bracket indicates six recent papers, listed in alphabetical order by author, that cite the Kandel and Schwartz reference. The list includes, for example, the following item: ICHINOSE, M. BRAIN RES 549 146 91.

This tells us that a paper citing the Kandel and Schwartz (1982) article was published in 1991 by M. Ichinose in volume 549 of the journal *Brain Research,* beginning on page 146. A paper that cites your key reference in its bibliography is probably appropriate to your topic and therefore worth consulting. If you wished to have more information before going to find volume 549 of *Brain Research*, you could look up the complete citation of the paper, including its full title, in the Source Index

KANDA M VOL PG YR
88 JPN J GENET 63 127
CHEN LFO BOTAN B A S 32 153 91
89 ANN PROBAB 17 379
KANDA M NAG MATH J 122 63 91 N
89 J BIOCHEM-TOKYO 105 653
SCHOLTEN JD SCIENCE 253 182 91
89 SINGAPORE PROB C
KANDA M NAG MATH J 122 63 91 N
KANDA N
83 P NATL ACAD SCI USA 80 4069
HASUMI K BIOC BIOP A 1083 289 91
87 CANCER RES 47 3291
HIRANO Y FEBS LETTER 284 235 91
KANDA P
80 J LIPID RES 21 257
SONNET PE CHEM PHYS L 58 35 91
KANDA S
67 J POLYM SCI C 17 151
ASWAR AS COLLOID P S 269 547 91
81 J PHYSIOL-LONDON 299 127
STERIN ABF ARCH I PHYS 99 141 91
81 PHYS CHEM BIOL 25 167
AZUMA Y CLIN CHEM 37 1132 91 L
KANDA T
** UNPUB J PROPULSION P
KANDA T J PROPUL P 7 431 91
79 MOSQ NEWS 39 568
FOLEY DH J AUS ENT S 30 109 91
83 ACTA OTOLARYNGOL S S 393 25
YOSHIHAR T ACT OTO-LAR 111 607 91
86 BIOCHEMISTRY-US 25 1159
KITAHATA S BIOCHEM 30 6769 91
86 J GEN APPL MICROBIOL 32 541
BINNINGE DM MOL G GENE 227 245 91
87 EAR RES JPN 18 501
KANDA T ACT OTO-LAR 97 91
87 J VIROL 62 610
87 JPN J CANCER RES 78 103
PHELPS WC CURR T MICR 144 153 89 R
88 J VIROL 62 610
FIRZLAFF JM P NAS US 88 5187 91
HASHIDA T J GEN VIROL 72 1569 91
IFTNER T CURR T MICR 144 167 89 R
KLEINHEI A " 144 175 89 R
WILBUR DC YALE J BIOL 64 113 91
88 VIROLOGY 165 321
MUNGER K J VIROLOGY 65 3943 91 N
SCHEFFNE M P NAS US 88 5523 91
TAKAMI Y INT J CANC 48 516 91
89 CLIN NEUROPATHOL 8 34
FUKUTANI Y J NEUROL 238 191 91
89 CLIN NEUROPATHOL 8 134
KANDA T J NEUR SCI 103 42 91
89 MOL GEN GENET 216 526
HASEBE K MYCOLOGIA 83 354 91
89 NAL TR1002 NAT AER L
KANDA T J PROPUL P 7 431 91
91 IN PRESS BRAIN 114
KANDA T J NEUR SCI 103 42 91
KANDA Y
74 J BIOL CHEM 249 6796
HARMS PJ COMP BIOC B 99 239 91
82 IEEE T ELECTRON DEV 29 64
DONZIER E SENS ACTU-A 26 357 91
82 IEEE T ELECTRON DEV 29 151
MUNTER PJA SENS ACTU-A 27 747 91
86 METALLOGRAPHY 19 461
TAKAO Y J NUCL MAT 179 298 91
87 JPN J APPL PHYS PT 1 26 1031
CHUNG GS ELECTR LETT 27 1098 91
88 ANAL CHIM ACTA 207 269
VANVEEN EH ANALYT CHE 63 1441 91
90 ANAL CHEM 62 2084
FOX DL ANALYT CHE 63 R292 91
KANDA Z
76 CHEM LETT 199
78 INORG CHEM 17 910
OCONNOR JM J AM CHEM S 113 4530 91
KANDALA CVK
87 T ASAE 30 793
88 INT AGROPHYSICS 4 3
88 T ASAE 31 1890
NELSON SO T ASAE 34 507 91
89 J AGR ENG RES 44 125
NELSON SO T ASAE 34 507 91
" " 34 513 91
KANDALAF N
82 JAMA-J AM MED ASSOC 248 2166
SOTANIEM KA J NE NE PSY 54 645 91 N
KANDALL S
77 EARLY HUM DEV 1 1
WITTMANN BK INT J ADDIC 26 213 91
KANDALL SR
75 ADDICT DIS INT 2 347
WITTMANN BK INT J ADDIC 26 213 91
77 EARLY HUM DEV 1 159
DOBERCZA TM J PEDIAT 118 933 91
83 AM J DIS CHILD 137 378
DOBERCZA TM J PEDIAT 118 933 91
HOEGERMA G CLIN PERIN 18 51 91
KANDARAKIS ED
85 NEUROPEPTIDES 6 21
JARVINEN A PHARM TOX 68 371 91
KANDASAMY K
** IN PRESS Z PHYS CHEM
SAKAMOTO Y SCR MET MA 25 1629 91
88 Z PHYS CHEM 158 253
89 Z PHYS CHEM NEUE FOL 163 41

KANDEL DB VOL PG YR
86 ARCH GEN PSYCHIAT 43 255
MOREAU D J AM A CHIL 30 642 91
86 ARCH GEN PSYCHIAT 43 746
WALTER HJ J AM A CHIL 30 556 91
87 INT J ADDICT 22 319
GREEN G BR J ADDICT 86 745 91
KANDEL E
78 VOPR NEIROKHIR 1 51
SALEH J NEUROSURGE 29 113 91
91 IN PRESS COLD SPRING 55
EDELMAN GM ANN R BIOCH 60 155 91 R
KANDEL EI
77 J NEUROSURG 46 12
SISTI MB J NEUROSUR 75 40 91
KANDEL ER
61 J NEUROPHYSIOL 24 225
DOZE VA NEURON 6 889 91
ISOKAWA M BRAIN RES 551 94 91
61 J NEUROPHYSIOL 24 243
ALBOWITZ B EUR J NEURO 3 570 91
DAVENPOR R BIOL CYBERN 65 47 91
LACROIX D J PHARM EXP 257 1081 91
66 J PHYSIOL-LONDON 183 287
FERGUSON GP J EXP BIOL 158 63 91
68 ELECTROPHYSIOLOGICAL 385
72 RES PUBLICATIONS ASS 50 91
KUPFERMA I PHYSIOL REV 71 683 91 R
76 CELLULAR BASIS BEHAV
ARSHAVSK YI DAN SSSR 318 232 91
COLEBROO E CELL MOL N 11 305 91
FERGUSON GP J EXP BIOL 158 63 91
" " 158 97 91
VENABLE N BEHAV PHAR 2 161 91
79 BEHAVIORAL BIOL APLY
FERGUSON GP J EXP BIOL 158 63 91
" " 158 97 91
79 HARVEY LECT 73 19
ISEKI M JPN J A P 1 30 1117 91
79 HARVEY LECTURES SERI 73 19
SAKHAROV DA J EVOL BIOC 26 557 90 D
80 CELLULAR BASIS BEHAV
ARSHAVSK YI NEUROPHYSIO 22 579 90
" " 22 587 90
KUDRASHO IV " 22 525 90
81 NATURE 293 697
RICHES IP J NEUROSC 11 1763 91
81 PRINCIPLES NEURAL SC
FRIEDMAN R J AM GER SO 39 650 91
81 PRINCIPLES NEURAL SC 731
STELLA A PHYSIOL REV 71 659 91 R
82 SCIENCE 218 433
ASZODI A P NAS US 88 5832 91
BUCHNER E J NEUROGEN 7 153 91 R
FABER DS BRAIN BEHAV 37 286 91
ICHINOSE M BRAIN RES 549 146 91 N
MASSICOT G NEUROSCI B 15 415 91 R
SCHWARTZ JH BIOCH SOC T 19 387 91
83 AM J PSYCHIAT 140 1277
HARRISON PJ J NERV MENT 179 309 91
85 PRINCIPLES NEURAL SC
JULESZ B REV M PHYS 63 735 91 R
MUSILA M BIOSYSTEMS 25 179 91
85 PRINCIPLES NEURAL SC 132
TAKAHASH M BRAIN RES 551 279 91
86 FIDIA RES F NEUROSCI 7
BRAVAREN NI ZH VYSS NER 41 107 91
86 NEUROSCIENCE AWARD L 7
ARSHAVSK YI DAN SSSR 318 232 91
86 NEUROSCI RES 3 498
GARCIAGI M COMP BIOC B 99 65 91
MASSICOT G NEUROSCI B 15 415 91 R
87 SYNAPTIC FUNCTION 471
BUCHNER E J NEUROGEN 7 153 91 R
89 J NEUROPSYCHIAT 1 103
HARRISON PJ J NERV MENT 179 309 91
89 PRINCIPLES NEUROSCIE 620
VRENSEN GFJ EXP EYE RES 52 647 91
KANDEL G
83 ARCH INTERN MED 143 1400
JAIN A CHEST 99 1403 91
SUMIMOTO T AM HEART J 122 27 91
KANDEL J
74 J BIOL CHEM 249 2088
LOBBAN MD BIOC BIOP A 1078 155 91
SEABROOK RN EUR J BIOCH 198 741 91
KANDEL M
78 J BIOL CHEM 252 679
SILVERMA DN CAN J BOTAN 69 1070 91
78 J BIOL CHEM 253 679
HUSIC HD CAN J BOTAN 69 1079 91
KANDEL RA
87 BIOCHEM INT 15 1021
90 BIOCHIM BIOPHYS ACTA 1053 130
CRUZ TF BIOCHEM J 277 327 91
90 J RHEUMATOL 17 953
BALBLANC JC REV RHUM 58 343 91
CRUZ TF BIOCHEM J 277 327 91
DEAN DD SEM ARTH R 20 2 91
KANDEL SN
73 CLIN ORTHOP RELAT R 96 108
SMYTH GB CORNELL VET 81 239 91
KANDELAKI HS
87 BIOCHIM BIOPHYS ACTA 170 893
AGOSTIAN A J PHOTOCH A 58 201 91
KANDELAKI TS
73 MIKROELEKTRONIKA 2 259
VLASOV SI IVUZ FIZ 34 121 91 N
KANDELER E

KANDIL A VO
SOLIMAN FM PHOSPHOR S 6
KANDIL AT
75 J INORG NUCL CHEM 37 229
DUKOV IL ACT CHIM HU 12
SAAD EA MICROCHEM J 4
80 J INORG NUCL CHEM 42 149
DUKOV IL ACT CHIM HU 12
KANDIL E
70 BRIT J DERMATOL 83 405
FETSCH JF MOD PATHOL
KANDIL OM
87 FED PROC 46 441
AQEL MB J ETHNOPHAR 3
KANDIL SH
86 J MATER SCI LETT 5 112
AFIFY N J NON-CRYST 12
KASSEM ME J THERM ANA 3
KANDILE NG
** IN PRESS ACTA CHIM H
** UNPUB ACTA CHIM HUNG
ISMAIL MF ACT CHIM HU 1
KANDILOV NK
63 IZVESTIYA AKADEM BMN 6
LANDSBER JH SYST PARAS 1
KANDIYOTI R
84 FUEL 63 1583
IBARRA JV FUEL 7
KANDLER O
01 BERGEYS MANUAL SYSTE 2 1
TOBA T LETT APPL M 1
68 J BACTERIOL 96 1935
SIJTSMA L NETH MILK D
80 BIOCH PLANTS 3 221
STAHL B ANALYT CHE
82 ENCY PLANT PHYSL A 13 348
CRESPI MD PLANT PHYSL 5
83 A VAN LEEUW J MICROB 49 2
84 BERGEYS MANUAL SYSTE 2 1
TSENG CP J BACT 17
84 NATURWISSENSCHAFTLIC 37 9
SEAWARD MRD LICHENOLOGI 2
85 BACTERIA TREATISE ST 7 41
JOHANNSE L INFEC IMMUN 5
85 KORTZFLEISCH 20
KANDLER O WATER A S P
86 ARCHAEBACTERIA 85 198
KOCH AL FEMS MIC R
86 BERGEYS MANUAL SYSTE 1
BUNCIC S J FOOD PROT 5
86 BERGEYS MANUAL SYSTE 2 1
JONSSON E J APPL BACT 7
KENNES C APPL MICR B 3
MOLLET B J BACT 17
86 BERGEYS MANUAL SYSTE 2 12
NGUYENTH C J APPL BACT 7
PILONE GJ AM J ENOL V 4
POSNO M APPL ENVIR 5
87 AFZ 42 715
EVERS FH WATER A S P
KANDLER O "
MONCHAUX P "
88 ALLG FORST JAGDZTG 159 179
KANDLER O WATER A S P 5
89 12 INT S MOL BEAMS
KANDLER O Z PHYS D 1
89 12 INT S MOL BGEAMS
LEISNER T Z PHYS D 20
89 1 P STAT SEM PBWU FO 7
KANDLER O WATER A S P 54
90 THESIS U KONSTANZ
LEISNER T Z PHYS D 20
KANDLER P
1862 CODICE DIPLOMATICO I
DEGASPER C TECTONOPHY 19
86 BERGEYS MANUAL SYSTE 1
LEBRAS G EUR J BIOCH 1
KANDLER R
26 WISS MEERESUNTERS KO 16
GUSTAFSO RG AQUACULTUR 9
90 LANCET 335 469
DHUNA A NEUROLOGY 4
KANDOLF R
85 J MOL CELL CARDIOL 17 167
85 P NATL ACAD SCI USA 82 4818
REIMANN BY J VIROLOGY 6
87 CIRCULATION 76 262
MATOBA Y JPN CIRC J 5
88 COXSACKIEVIRUSES GEN 29
89 CONCEPTS VIRAL PATHO 3 28
89 SPRINGER SEMIN IMMUN 11
REIMANN BY J VIROLOGY 65
90 NEW ASPECTS POSITIVE 340
ROSE NR IMMUNOL TO 12
KANDOLO K
85 J CLIN MICROBIOL 21 980
MCGAREY DJ EXPERIENTIA 4
KANDOOLE BF
** ADMARCS COST STRUCTU
SAHN DE FOOD POLICY 1
KANDORI H
** UNPUB
KOBAYASH T CHEM P LETT 18
89 BIOCHEMISTRY-US 28 6460
89 BIOPHYS J 56 453
89 PHOTOCHEM PHOTOBIOL 49 181
SHICHIDA Y BIOCHEM 30
KANDORI K

FIGURE 6.

Detail of a page of the Citation volume from the 1991 *Science Citation Index.* The lower arrow indicates the key paper on which our search is based, a 1982 paper in *Science* (volume 218, beginning on page 433).

volume of *Science Citation Index.*

Looking elsewhere in Figure 6, note that most journals are not listed by their full names. To avoid confusion, each volume of the *Science Citation Index* begins with an explanation of all journal abbreviations used. You will find, for example, that IMM CELL B stands for *Immunology and Cell Biology*, while CAN J MICRO stands for the *Canadian Journal of Microbiology. Science Citation Index* is published at two-month intervals and is consolidated into a smaller number of volumes yearly. It is available on CD-ROM and electronically in the UK via the BIDS service.

Using *Zoological Record*

Another useful indexing service is the *Zoological Record*, published jointly

Subject	Page
Nucleolus	313
Nucleoplasm	313
Nucleus	313
Number of generations	390
Nutrition	350
Obituaries	291
Observation techniques	298
Oceanic islands	648
Ohio	637
Oklahoma	637
Oligocene	655
Oman	569
Omnivorous feeding	335
Ontario	619
Ontogenesis	428
Oogenesis	386
Optomotor response	495
Orange Free State	606
Oregon	637
Organelles	313
Organic chemicals	372
Organized study	271
Oriental region	596
Orientation	509
Origin of taxon	433
Osmotic relations	363
Ossification	318
Ovarian cycle	387
Ovary	387
Overpopulation & culling	284
Overwintering	520
Oxidative metabolism	370
Oxygen consumption	362
Plant caused disorders	538
Plasma	358
Play	504
Pleistocene	654
Pliocene	654
Plumage	314
Poland	569
Pollination	526
Pollutant content	368
Pollutant metabolism	370
→ **Pollution**	277 ←
Polygamy	411
Polymorphism	437
Pond	472
Popular works	292
Population censuses	464
Population changes	455
Population density	452
Population density measurement	299
Population dynamics	449
Population genetics	432
Population genetics techniques	298
Population inbreeding	432
Population mapping	299
Population regulation	460
Population sampling	299
Population sex ratio	447
Population size	450
Population structure	446
Population study	445
Portugal	569
Postembryonic development	428
Posture	508
Precopulatory behaviour	414

FIGURE 7.

Detail of a page of subject listings from the 1990 Chordata volume of the *Zoological Record.* The arrow indicates our topic of interest, directing us to page 277.

by BIOSIS and the Zoological Society of London. *Zoological Record* is published once a year in several volumes. Each is devoted to a particular animal phylum or group of related phyla, for example, the Mollusca, the Annelida and the Chordata. At the front of each volume is a section arranged by subject, as shown in the example in Figure 7 taken from the 1990 volume covering the members of our own phylum, the Chordata. It is also available in electronic form.

Suppose we wish to find references on the influence of acid rain on the reproductive success of birds. As indicated by the arrows in Figure 7, the subject of Pollution is probably our best bet, so we turn to page 277 of the same volume. On page 277, we find a variety of interesting pollution-related references (see the circled heading in Figure 8), most of which were published in either 1989 or 1990, the time-span covered by this particular volume of *Zoological Record*. Under the heading *Chemical Pollution*, a paper by Drent *et al.* (see arrow) seems particularly promising and we are referred to reference no. 2690. The complete reference for this paper, including the names of all co-authors, the title of the paper, and the complete page numbers, is given in a separate section towards the front of the volume, as shown in Figure 9 (page 44). Note the arrow in front of the citation for the Drent *et al.* article. These references are listed alphabetically and also numerically, so that we could have looked up the reference under 'Drent' rather than by number.

In addition to the author and subject indices already discussed, each volume of *Zoological Record* contains a Geographical Index (in which references are arranged by geographical area and country), a Systematic Index (arranged by taxonomic group) and a Paleontological Index covering animals known only as fossils (arranged by geological epoch and era).

Using *Biological Abstracts*

Another widely used service is *Biological Abstracts*, published by BIOSIS in Philadelphia. It can be frustrating to use the subject index in the printed form. You begin in a straightforward manner, by looking up key words relevant to the topic being researched. Suppose, for example, you are looking for papers discussing the molecular basis for the action of

Zoological Record, Vol. 126, 1989/90

Population changes relationships, Pennsylvania
GOODRICH, L.J., ET AL (3610)

POLLUTION

Abundance/biomass comparison indications, Netherlands (marine)
MEIRE, P.M., ET AL (6261)
Biological effects, England
Cygnus olor GLASER, G.A. (3515)
Breeding population decline relationship, lake, Northern Ireland
Melanitta nigra PARTRIDGE, J.K. (7096)
Compilation of entangled species in beach litter survey, West Germany
LIEDTKE, G., ET AL (5650)
Conservation implications, Turkey AKCAKAYA, H.R. (248)
Conservation role of RSPB, review, United Kingdom SAMSTAG, T. (8095)
Discarded fishing tackle, ropes & plastic litter
Mortality due to entrapment, seabirds, Netherlands
CAMPHUYSEN, C.J. (1656)
Habitat damage indicator & conservational significance, West Germany
Acrocephalus scirpaceus ANON (6)
Habitat utilization relationships, lake, Japan
Anatinae SUGIMORI, F., ET AL (8935)
Indicator, restoration aims, Great Lakes
Larus argentatus GILBERTSON, M. (3485)
Lake eutrophication, endangered status, West Germany
Tachybaptus ruficollis REICHHOLF, J. (7713)
Marine habitat, indicator role BATTY, L. (712)
Plastic debris incorporation into nests, Maine
Phalacrocorax auritus PODOLSKY, R.H., ET AL (7408)
Review, North America LEIGHTON, F.A. (5554)
Wetland population changes, indicator value, Michigan
MAINONE, R., ET AL (5905)

CHEMICAL POLLUTION

Accidental poisoning with fenthion, Orange Free State
Buteo buteo COLAHAN, B.D., ET AL (2023)
Acid rain effect on forest & breeding population changes, W Germany
OELKE, H. (6896)
Acid rain, effects on aquatic species, Europe & North America
DIAMOND, A.W. (2545)
→ Acid rain, relations with egg shell inferior quality, forest, Netherlands
Parus major DRENT, P.J., ET AL (2690)
Acute lead poisoning effect on tissue lead levels & blood values
Cygnus olor O'HALLORAN, J., ET AL (6877)
Agrichemicals in prairie wetlands potential effects & management, Canada
Anatidae FORSYTH, D.J. (3200)
Agrichemicals in prairie wetlands potential effects & management, USA
Anatidae GRUE, C.E., ET AL (3808)
Biological effects, past, present & future, overview
Falco peregrinus RATCLIFFE, D.A. (7652)
Blood protoporphyrin levels use as indicator of lead poisoning
Callonetta leucophrys PASSER, E.L., ET AL (7105)
Breeding population change relationships, Switzerland
Falco tinnunculus KAESER, G., ET AL (4929)
Tyto alba KAESER, G., ET AL (4929)

FIGURE 8.

Detail of page 277 from *Zoological Record,* showing references on chemical pollution.

Drake, O.F. (2686)
Continental starlings in Devon.
Devon Birds **41**(4) 1988: 73-75, illustr. [In English]

Dransfeld, H. *see* Bangjord, G.

Dranzoa, C. & Rodrigues, R. (2687)
Two new records for Uganda.
Scopus **14**(1) 1990: 32-33. [In English]

Draulans, D. (2688)
Timing of breeding and nesting success of raptors in a newly colonized area in north-east Belgium.
Gerfaut **78**(4) 1988: 415-420, illustr. [In English with Dutch & French summaries]

Draulans, D. & van Vessem, J. (2689)
Some aspects of population dynamics and habitat choice of grey herons (*Ardea cinerea*) in fish-pond areas.
Gerfaut 77(4) 1987: 389-404, illustr. [In English with Dutch & French summaries]

Dreissens, G. *see* van der Burg, E.

→ **Drent, P.J. & Woldendorp, J.W.** (2690)
Acid rain and eggshells.
Nature (Lond) **339** No 6224 1989: 431, illustr. [In English]

Drent, R. & Klaassen, M. (2691)
Energetics of avian growth: the causal link with BMR and metabolic scope.
NATO Adv Study Inst Ser Ser A Life Sci **173** 1989: 349-359, illustr. [In English]

Drent, R.H. *see* de Boer, W.F.
Drewes, L.A. *see* Flammer, K.

Drgonova, N. & Janiga, M. (2692)
Nest structure of alpine accentors (*Prunella collaris*) (Scop., 1769) in the low Tatras.
Biologia (Bratisl) **44**(10) 1989: 983-993, illustr.
[In English with Russian & Slovak summaries]

Drickamer, L.C. (2693)
Pheromones: behavioral and biochemical aspects.
Adv Comp Environ Physiol **3** 1989: 269-348, illustr. [In English]

Driel, F. van *see* van den Bergh, L.M.J.
Driesch, A. von den *see* Boessneck, J.
Driessens, G. *see* Buys, P.; van der Burg, E.

Drillon, V. (2694)
Analyse des causes de regression du grand tetras dans le massif de la Haute Meurthe.
Ciconia **13**(1-2) 1989: 11-18, illustr. [In French with English summary]

Drimmelen, B. van *see* Munro, W.T.

FIGURE 9.

Detail of a page from the author index at the front of *Zoological Record*, with an arrow indicating the complete reference for the paper by Drent and Woldendorp.

prolactin, the pituitary hormone that triggers milk production in mammals. Key words, such as *prolactin*, are listed in the central portion of each column as shown in Figure 10.

Bio

Subject Context	▼ Keyword		Ref. No.
VICAL GANGLION NEURONS		TO A SPECIFIC TARGET SYMP	98677
MAGNOCELLULAR NEURONS		TO THE NEUROHYPOPHYSIS I	98533
EURONS AND NERVE FIBERS		TO THE RAT SUBMANDIBULA	98615
INTIMAL TEAR ULCER-LIKE	**PROJECTION**	/CLINICAL ASSESSMENT OF R	92887
T OF THE NORADRENERGIC		FROM LOCUS COERULEUS TO	98525
F THE OLFACTO-RETINALIS		IN COHO SALMON ONCORHYN	101991
IN THE TECTO-GENICULATE		IN THE TREE SHREW TUPAIA-	98739
ONPYRAMIDAL ENTORHINAL		NEURONS CONTRIBUTING TO	98876
C CEREBELLAR AND SPINAL		NEURONS IN RAT TRIGEMINA	98172
ENTIFIED UNIGLOMERULAR		NEURONS IN THE ANTENNAL	96558
L AXONS MACACA-MULATTA		NEURONS LOCAL CIRCUIT NE	98740
LGORITHM/ FILTERED BACK		ON SHARED-MEMORY MULTIP	97318
RESOLUTION TEST TARGET		OPHTHALMOSCOPE HUMAN P	101838
THE RABBIT DIFFERENTIAL	**PROJECTIONS**	AFFERENT PROJECTIONS CE	98168
RGE PATTERNS AND SPINAL		CAT COMPARATIVE PHYSIOL	98856
AL PROJECTIONS AFFERENT		CEREBRAL PEDUNCLE SUBST	98168
IGRAL AND NIGROSTRIATAL		IN THE MACAQUE CAUDATE	98638
IGIN OF TRIGEMINOSPINAL		IN THE RAT HORSERADISH	98614
NIN-LIKE IMMUNOREACTIVE		IN THE TELEOST FISH POEC	98640
E 5 HYDROXYTRYPTAMINE/		OF NEURONS IN THE VENTR	98789
ALYX OF HELD HISTOLOGY/		OF PHYSIOLOGICALLY CHAR	98526
ABELING HISTOLOGY/ DUAL		OF SECONDARY VESTIBULAR	98578
RADE TRACING HISTOLOGY/		OF THE PARAVENTRICULAR	98178
NG TECHNIQUES/ EFFERENT		OF THE PERIAQUEDUCTAL G	98168
DING CATECHOLAMINERGIC		RAT CHOLINERGIC NEURON	98644
RIGIN OF SEROTONINERGIC		TO DIFFERENT REGIONS OF	98166
LABEL/ ORIGIN OF SPINAL		TO THE ANTERIOR AND POS	98188
TANCE P-IMMUNOREACTIVE		TO THE PARAVENTRICULAR	98644
GES HAVING UNIQUE HEAD		TP1 VP3 VP6 POLYPEPTIDE	102706
EIN ANGIOGRAM ENLARGER	**PROJECTOR**	/TRANSPARENCY OVERLAY A R	101889
ER OF A LARGE FAMILY OF	**PROKARYOTE**	AND EUKARYOTE SURFACE T	100032
TIC TRANSFORMATION OF A		BACILLUS-MEGATERIUM PLA	95133
EQUIVALENTS FROM OTHER	**PROKARYOTES**	OR EUKARYOTES WITHOUT	93581
OF HYDROGEN PEROXIDE IN	**PROKARYOTIC**	AND EUKARYOTIC ALGAE AC	100517
OT GEL ELECTROPHORESIS/		METALLOTHIONEIN GENE C	95711
POLE WESTERN AUSTRALIA		MICROORGANISM BIOSTRA	98986
ON TO ALL RHO-DEPENDENT		TRANSCRIPTION TERMINAT	95195
RESSIN GROWTH HORMONE	**PROLACTIN**	/ASSESSMENT OF HYPOTHALA	98439
OREPINEPHRINE DOPAMINE		/HYPOTHALAMIC SITES OF CAT	98708
GROWTH RATE MILK YIELD		/LONG-TERM EFFECT OF A SHE	94614
UMORS LIGHT MICROSCOPY		/MAGNETIC RESONANCE IMAGI	97710
GING ON SERUM LEVELS OF		AND ALPHA MELANOTROPIN IN	94597
FLUOROSIS ON PITUITARY		AND HYPOTHALAMIC 5-HT CON	102387
NNUAL RHYTHM OF PLASMA		CONCENTRATIONS IN EWES EX	94632
E PULSATILE RELEASE FSH		CORTISOL/ ENDURANCE TRAIN	94499
ON BEHAVIOR/ EXOGENOUS		DELAYS PHOTO-INDUCED SEXU	94629
RIAN STEROID RECEPTORS/		ENHANCES UTEROGLOBIN GENE	101515
Y/ ABSENCE OF PITUITARY		EPITOPES IN IMMUNOREACTIV	94617
DIOIMMUNOASSAY FOR EEL		FRESHWATER ADAPTATION SEA	94647
TO GROWTH HORMONE AND		FROM ATLANTIC COD GADUS-M	94605
VELS TRH/ EXPRESSION OF		GENE IN INCUBATING HENS M	100580
OR PITUITARY IN CULTURE		GONADOTROPIN THYROTROPIN	94635
RAT BROMODEOXYURIDINE		GROWTH HORMONE ACTH LUTE	94643
S HUMAN NOREPINEPHRINE		GROWTH HORMONE CORTISOL	94581
IN 5 HYDROXYTRYPTAMINE		GROWTH HORMONE DEPRESSIO	100753
HREE DIFFERENT METHODS		GROWTH HORMONE HISTOLOG	94645
RT OF SEVEN CASES HUMAN		GROWTH HORMONE TRH GONA	97438
NIZING HORMONE FSH TSH		GROWTH HORMONE/ THE IMPA	94641
MA ANTINEOPLASTIC-DRUG		IMMUNOTHERAPY/ A NEW APP	98106
UTEINIZING HORMONE AND		IN LACTATING AND NONLACTA	94608
NIZING HORMONE FSH AND		IN OVARIECTOMIZED PITUITAR	98442
ECTOMY/ INVOLVEMENT OF		IN THE REGULATION OF PLASM	94668
PIN SECRETION AND BASAL		LEVELS DURING DOPAMINE D-1	99849
NEOPLASTIC-DRUG/ SERUM		LEVELS IN PATIENTS TREATED	98029
LEX AROMATASE ESTROGEN		LUTEINIZING HORMONE FSH/	101561
SE IN THE INFANTILE RAT		LUTEINIZING HORMONE/ SEXU	94639
MAN EPILEPSY/ CAPILLARY		MEASUREMENT FOR DIAGNOSI	98225
EWE RADIOIMMUNOASSAY/		MESSENGER RNA CONCENTRATI	101565
OPES IN IMMUNOREACTIVE		OF RAT BRAIN MONOCLONAL A	94617
MULATION/ INTRACRANIAL		PERFUSION INDUCES INCUBAT	94630
UTEINIZING HORMONE FSH		PLASMA CONCENTRATIONS/ OP	91827
TENEDIONE TESTOSTERONE		PLASMA LEVELS FOLLICULAR F	99446
YLATE DOPAMINE AGONIST		PROGESTERONE OVULATION/ E	101590
AND SHORT FORMS OF THE		→ RECEPTOR ON PROLACTIN-IND	95343 ←
/ DYNAMICS OF PULSATILE		RELEASE DURING THE POSTPAR	101553
EFFECTS OF DOPAMINE ON		RELEASE FROM ECTOPIC AND H	94640
OPIN RELEASING HORMONE		REPRODUCTION/ ENDOCRINE C	91836
ATING SCALE STATISTICS/		RESPONSE TO SUBMAXIMAL ST	100727
EM/ KALLIDIN STIMULATES		SECRETION FROM INDIVIDUAL	94615
IATED CORTISOL ACTH AND		SECRETION IN HUMANS PHAR	98707
S/ GROWTH HORMONE AND		SECRETION IN HYPOPHYSIAL S	94596

Subject Context	▼ Keyword	
ED LYMPHOCYTE REACTION		CYTC
IS RABBIT SMOOTH MUSCLE		DED!
ROSODIMETHYLAMINE CELL		DELA
CULTURE EXTRACAPILLARY		FIBR
S HUMAN HEPATOCYTE CELL		GENE
STITUTIVE PROMOTER CELL		GENE
NTERMEDIATES PHAGOCYTE		GRAN
CER WITH A HIGH RATE OF		HUM
ALS IN PROMOTING T CELL		HUM
LOGIC AGENT LYMPHOCYTE		HUM
GE CHANGES AND SYNOVIAL		IN C
LTERATION OF LYMPHOCYTE		IN M
HORMONE TSH DNA/ CELL		IN PI
IS IN RAT PANCREAS CELL		INHIE
PHARMACOKINETICS T CELL		INHII
HOKINE SYNTHESIS T CELL		LYMP
PSORIASIS KERATINOCYTE		MAL'
LYMPHOCYTES HUMAN CELL		MAR
CHOLESTEROL LYMPHOCYTE		MAS
DEFICIENCY VIRUS T CELL		MEN
-STIMULATION HEPATOCYTE		MES
ONSE CELL ADHESION CELL		MIC
RTITION RAT PEROXISOMAL		MIC
URAL SUPPLY NERVE FIBER		NER'
S/ HYPERINSULINEMIA AND		OF E
OF PROTEIN KINASE C IN		OF C
ANGIOGENESIS FACTOR ON		OF E
ACTIVATION OF IN-VITRO		OF H
D WITH THE INDUCTION OF		OF N
YLASE IN THE CONTROL OF		OF T
L ATCC CCL64 CELLS CELL		PREI
PTASE ACTIVITY CELLULAR		RATI
STS AND HELA CELLS GENE		REGL
FERASE BETA GLOBIN CELL		S PH
YTES MESSENGER RNA CELL		SHAF
ATORY PROCESS CELLULAR		STIM
ARCH METABOLISM CALLUS		STRU
VE IMMUNITY LYMPHOCYTE		THRO
SYNTHESIS BOMBESIN CELL		TWO-
TEINASE CATHEPSIN L AND	**PROLIFERATION-ASS**	
ELLULAR DNA CONTENT AND	**PROLIFERATIVE**	ACTIV
REDUCTION PRETREATMENT		ACTIV
ATIVE VITREORETINOPATHY		DIAB
ISTRY/ VITRONECTIN AND		INTR
PATHY/ VITRONECTIN AND		INTR
TERACTIONS IN MESANGIAL		NEPH
ROGENITORS WITH A HIGH		POTE
TIVE TO THE TOXIN MOUSE		RESP
TA T LYMPHOCYTES HUMAN		RESP
AR CELL MITOGEN INDUCED		RESP
D-TYPE HYPERSENSITIVITY		RESP
160-SPECIFIC LYMPHOCYTE		RESP
D NK CELLS CYTOTOXICITY		RESP
IABETIC BACKGROUND AND		RETI
CLASS II-POSITIVE CELLS		T-LY
NES FROM PATIENTS WITH		VITR
MAN RETINAL DETACHMENT		VITR
N. VARIETY 85-59 CALLUS	**PROLINE**	ACCUMULAT
ENDIPITOUS FINDING OF L		AS A PRECU
CARBOXYLASE EC 4.1.1.31		ENZYME STR
NEURAMINIC ACID GLYCINE		HISTAMINE/
RO ORGAN CULTURE MODEL		HYDROXYPRO
OF SALSOLA-SODA ENZYMES		INCOMPATIBI
MA-MANSONI HIGHER FREE		LEVELS IN T
ULATION AND EFFECT OF L		ON SOMATIC
TED MICE EGG GRANULOMA		OXIDASE EC
SEQUENCE DATA/ ROLE OF		RESIDUES IN
R DEMONSTRATION THAT 9		SUBSTANCE
CHROMATOGRAPHY OF THE	**PROLINE-CONTAINI**	
Y-7-METHYLNONANE INSTAR	**PROLONGATION**	DEAT
DIOGRAPHY/ QT INTERVAL		IN T
ASSESSEMENT FOR AN ORAL	**PROLONGED-RELEAS**	
LONING OF PORCINE BRAIN	**PROLYL**	ENDOPEPTIDAS
→ PARENTLY IDENTICAL WITH		ENDOPEPTIDAS
F THE OXYTOCIN FRAGMENT	**PROLYL-LEUCYL-GL**	
A SYNTHESIS/ FAILURE OF	**PROLYLPEPTIDYL**	ISO
ROMASTIGOTE METACYCLIC	**PROMASTIGOTE**	LIPO
IGOTE LOGARITHMIC PHASE		MET
YTE PROLIFERATION ASSAY	**PROMASTIGOTES**	CU
INTRAVENOUS DRUG ABUSE	**PROMISCUOUS**	SEX/
FFERENTIATION OF HUMAN	**PROMONOCYTIC**	U93

FIGURE 10.

Detail of a page from the April-May 1991 volume of *Biological Abstracts* listing the key word PROLACTIN and related references. The horizontal arrows target one reference of particular interest.

genes to be express[illegible] cells. The enhan[illegible] localized to a 200-[illegible] Rsa I restriction fragments, which contains sequence motifs similar to those found in the other T–cell receptor enhancers but not in the immunoglobulin enhancers.

95342. LETSOU, ANTHEA, SHERRY ALEXANDER, KIM ORTH and STEVEN A. WASSERMAN. (Dep. Biochem., Univ. Tex. Southwestern Med. Cent., Dallas, Tex. 75235.) PROC NATL ACAD SCI U S A 88(3): 810–814. 1991. **Genetic and molecular characterization of tube, a *Drosophila* gene maternally required for embryonic dorsoventral polarity.**—Loss of maternal function of the tube gene disrupts in signaling pathway required for pattern formation in *Drosophila*, causing cells throughout the embryo to adopt the fate normally reserved for those at the dorsal surface. Here we demonstrate that tube mutation also have a zygotic effect on pupal morphology and that this phenotype is shared by mutations in Toll and pelle, two genes with apparent intracellular roles in determining dorsoventral polarity. We [illegible]an describe the isolation of a functionally full–length tube cDNA identified in a phenotypic rescue assay assay. The tube mRNA is expressed maximally early in embryogenesis and again late in larval development, corresponding to required periods of tube activity as defined by distinct maternal and zygotic loss–of–function phenotypes in tube mutations. Sequence analysis of the cDNA indicates that the tube protein contains five copies of an eight–residue motif and shares no significant sequence similarity with known proteins. These results suggest that tube represents a class of protein active in signal transduction at two stages of development.

——►**95343.** LESUEUR, LAURENCE*, MARC EDERY*, SUDAH ALI, JACQUELINE PALY*, PAUL A. KELLY and JEAN DJIANE*. (Unite d'Endocrinol. Mol., Inst. Natl. de la Recherche Agronomique, 78352 Jouy–en–Josas Cedex, Fr.) PROC NATL ACAD SCI U S A 88(3): 824–828. 1991. **Comparison of long and short forms of the prolactin receptor on prolactin–induced milk protein gene transcription.**—The biological activities of long and short forms of the prolactin receptor have been compared. These two receptors expressed in mammalian cells were shown to bind prolactin with equal high affinity. The ability of these different forms to transduce the hormonal message was estimated by their capacity to stimulate transcription by using the promoter of a milk protein gene fused to the chloramphenicol acetyltransferase (CAT) coding sequences. Experiments were performed in serum–free conditions to avoid the effect of lactogenic factors present in serum. An $\approx$ 17–fold induction of CAT activity was obtained in the presence of prolactin when the long form of the prolactin receptor was expressed, whereas no induction was observed when the short form was expressed. The present results clearly establish that only the long form of the prolactin receptor is involved in milk protein gene transcription.

95344. LAU, C. K., M. SUBRAMANIAM, K. RASMUSSEN and T. C. SPELSBERG*. (Dep. Biochem. and Mol. Biol., Div. Endocrinol., Mayo Clin., Rochester, Minn. 55905.) PROC NATL ACAD SCI U S A 88(3): 829–833. 1991. **Rapid induction of the c–jun protooncogene in the avian oviduct by the antiestrogen**

FIGURE 11.

Detail of a page from *Biological Abstracts* giving complete citation for a paper by Lesueur *et al.*, along with a detailed summary (abstract) of that paper's contents.

Subject Context	▼ Keyword	Ref. No.
ANCE OF RACE 4 OF DOWNY	**MILDEW** DERIVED FROM INTERSPECIFIC CR	100288
/ RESISTANCE TO POWDERY	ERYSIPHE-GRAMINIS-F-SP-HORDEI	100266
/ RESISTANCE TO POWDERY	ERYSIPHE-GRAMINIS-F-SP-HORDEI	100267
AN 8 SACHEON 6 POWDERY	GROWTH EARLY MATURATION LOD	91732
S ON THE DEVELOPMENT OF	IN WINTER WHEAT ERYSIPHE-GRA	100146
EST LOCATIONS REGARDING	INFECTION ERYSIPHE-GRAMINIS D	100147
LUS PLANT FUNGUS DOWNY	MODELING AIR TEMPERATURE SOI	100172
RE/ SOURCES OF POWDERY	RESISTANCE IN WHEAT AND TRITI	100091
ITH A GENE FOR POWDERY	RESISTANCE PMD ERYSIPHE-CICHO	100291
PALDANGHOMIL POWDERY	RUST SCAB MORPHOLOGY LODGIN	91733
O THE HOSTS OF POWDERY	**MILDEWS** OF PAKISTAN LESPEDEZA-JUNCEA	92790
MINISTRATION/ EFFECT OF	**MILDRONATE** ON RAT HEART CONTRACTILI	99165
NXIETY DEPRESSION THREE	**MILE** ISLAND PENNSYLVANIA/ EFFECT OF	101371
BSTRUCTION/ CONGENITAL	**MILIARIA** CRYSTALLINA CHILD INTRAEPID	96454
OLOGOUS PROTEINS TO ITS	**MILIEU** ERWINIA-CAROTOVORA KLEBSIELLA	95185
HIA-CANIS INFECTIONS IN	**MILITARY** DOGS IN AFRICA AND REUNION	97038
S PSYCHOPATHOLOGY/ THE	FAMILY SYNDROME REVISITED B	100801
TUBE HUMAN OCCUPATION	MEDICINE GERMANY/ FUNCTION	101959
ONS EXPERIENCE FROM THE	OBSTACLE TRACK HUMAN OBSTA	99061
D MENTAL EDUCATION FOR	PRIMARY CARE PHYSICIANS HUM	100869
Y MANIPULATIVE BEHAVIOR	SYSTEM/ MALINGERING IN THE	100973
/ RESIDUES OF ETHION IN	**MILK** AFTER INTRAVENOUS ORAL AND DERM	102283
THE PRESENCE OF CALCIUM	ANALYSIS/ DIRECT SPECTROPHOTOME	94922
TION/ DETERMINATION OF	AND MAMMARY TISSUE CONCENTRATI	96976
TIFIED 2 PERCENT LOWFAT	AND SKIM MILK DAIRY PRODUCT DAI	94917
E ORIGIN OF A TRADITION	BOTTLE OPENING BY TITMICE AVES	92002
DDT RESIDUES IN BOVINE	BUTTER AND FAT IN DISTRICT CHAR	102286
LNEURAMINYLTRANSFERASE	CARBOHYDRATE-RICH DIET/ DIETARY	98899
XIDANTS SOYBEAN PROTEIN	CASEIN MAIZE GLUTEN ZEIN ALPHA	94907
ICAL CHANGES IN BUFFALO	CHEDDAR CHEESE/ EFFECT OF PROTE	94913
NE COMPOUNDS IN HUMAN	CONTAMINATION HEXACHLOROBENZEN	102272
N SORGHUM SUDAN GRASS	COW LIVESTOCK PROTEIN CONTENT D	91602
T LOWFAT MILK AND SKIM	DAIRY PRODUCT DAIRY INDUSTRY TR	94917
IN REPLETION ASSAY SKIM	DAIRY PRODUCT FOOD NUTRITION/ B	98937
OORGANISM DISINFECTANT	DAIRY PRODUCT FOOD PROCESSING/	93970
LIPASE IN HUMAN BREAST	DURING EXTENDED LACTATION LIPID	101574
RANES IN LAMBS FED COW	ENRICHED WITH SOLUBLE STARCH BE	91875
CTOR TGF-ALPHA IN HUMAN	EPIDERMAL GROWTH FACTOR COLOSTR	94475
GLYCOPROTEIN OF BOVINE	FAT GLOBULE MEMBRANE MAMMAL CO	91831
LEAD AND COPPER IN COW	FOOD CONTAMINATION ASTURIAS SPA	102296
M INFANT GIVEN MOTHER'S	FORTIFIED WITH PROTEIN FROM HUM	98948
ONCENTRATION OF HUMAN	GLUCOSE MAMMARY ALVEOLAR CELL L	101542
ALITY BACTERIAL QUALITY	GRADING SYSTEM/ EFFECTS OF STOR	94924
IN FROM HUMAN OR COW'S	GROWTH FORTIFICATION HIGH PERFO	98948
HEMICAL TREATMENTS COW	GUINEA-PIG ANAPHYLACTIC EFFECT	94914
ERBICIDES IN POOLED RAW	IN CONNECTICUT USA COW MAMMAL 2	102278
ICAL ASSESSMENT OF RAW	IN THE GDANSK REGION POLAND AND	97054
OD HUMAN WATER BREAST	INFANT FORMULA STATISTICS/ MEAS	98897
TISTICS/ MEASUREMENT OF	INTAKE TRACER-TO-INFANT DEUTERI	98897
OBACTERIACEAE ACIDOSIS/	LEVEL OF CHLORIDES AS AN INDEX	97075
MIC RETICULUM ORIGIN OF	LIPID GLOBULES OBTAINED USING L	101456
S OF MANGANESE CALCIUM	PHOSPHORUS COPPER AND ZINC HUMA	98929
ENT HEXAMETAPHOSPHATE	POWDER PROTEOLYSIS SEDIMENTATIO	94923
IA-MONOCYTOGENES FROM	POWDER USED FOR REFERENCE SAMPL	94868
WHEAT FLOUR CORN FIBER	POWDER/ FRACTAL STRUCTURE ANALY	94906
AMMAL LACTATION PERIOD	PRODUCTION ANIMAL BREEDING THER	91858
SS RESISTANCE PRIMIPARA	PRODUCTION GENETIC POTENTIAL AD	91852
OON PRODUCTION CUTTING	PRODUCTION KOREA/ A NEW PEARL M	91602
K YIELD IN DAIRY CATTLE	PRODUCTION SEX SIRE BIRTH WEIGH	91844
N BUTTERFAT PRODUCTION	PRODUCTION/ EFFECT OF FLAVOMYCI	91849
TION FEED INTAKE ENERGY	PRODUCTION/ PLASMA GROWTH HOR	91835
NKER TRAIT HERITABILITY	PRODUCTIVITY/ GENETIC POLYMORPH	91831
OR ON PROLACTIN-INDUCED	PROTEIN GENE TRANSCRIPTION OVIN	95343 ←
ENT UPON THE QUALITY OF	PROTEIN ISOLEUCINE THREONINE TR	94921
ROSCOPY/ MODULATION OF	PROTEIN SYNTHESIS THROUGH ALTER	101504
ATION OF MEAT CURD AND	PSEUDOMONAS BEEF CHICKEN FLUORO	94873
E CONDITION ON THE RAW	QUALITY BACTERIAL QUALITY MILK	94924
ICROBIAL CONTAMINATION	RANCIDITY/ EFFECT OF SELECTED A	94871
LACTATION MICROTUBULES	SECRETION IMMUNOCYTOCHEMISTRY/	101543
ING RAT MAMMARY GLAND	SECRETION STIMULATION 3 HYDROXY	94506
COW MILKING FREQUENCY	SECRETION/ AUTOCRINE REGULATION	101541
HERDS WITH A LOW BULK	SOMATIC CELL COUNT 1. DATA AND	96977
DING SUGARED CHOCOLATE	SUCROSE CALORIC CONSUMPTION PAL	98884
IN THE SYNTHESIS OF THE	SUGARS LACTOSE AND ALPHA-2 3 SI	101544
MYCOFLORA IN FERMENTED	TORULOPSIS-CANDIDA ASPERGILLUS-	94870
UTED CONCENTRATED SKIM	ULTRA-HIGH TEMPERATURE HEAT TRE	94923
OTHER FACTORS ON DAM'S	YIELD IN DAIRY CATTLE MILK PROD	91844
ARDIZATION OF LACTATION	YIELD IN NILI-RAVI BUFFALOES CA	91862
ENTAL FACTORS AFFECTING	YIELD IN PAKISTANI BUFFALOES BR	91861
NDUSTRY DAIRY INDUSTRY	YIELD LACTATION SEASONALITY/ PE	91854
N RESPONSE GROWTH RATE	YIELD PROLACTIN/ LONG-TERM EFFE	94614
ATIONS ON A PRODUCER OF	**MILK-COAGULATING** ENZYME COMPLEX B	94847
COPROTEIN IN GUINEA-PIG	**MILK-FAT-GLOBULE** MEMBRANE EVIDENC	101546
CULLING QUALIFICATIONS	**MILK-TESTING** PROCEDURES DAIRY PROD	91923
TIATION GOAT HEIFER COW	**MILKING** FREQUENCY MILK SECRETION/ AU	101541
IODOPHOR GERMICIDES IN	PARLOR UDDER WASH WATER SYS	93962
ED DRY COW MANAGEMENT	PROCEDURE EQUIPMENT PRODUCT	96977
AGE/ INFLUENCE OF PAPER	**MILL** EFFLUENT IRRIGATION ON THE POPU	102059

FIGURE 12.

Detail of page from *Biological Abstracts* showing that the Lesueur *et al.* reference could also have been located by using MILK as the key search word (see arrow).

MAJOR CONCEPT HEADINGS FOR ABSTRACTS

Use this list to locate the Abstracts that correspond to your research interest.

	Page
Aerospace and Underwater Biological Effects	AB-1
Agronomy	AB-1
Allergy	AB-28
Anatomy and Histology, General and Comparative	AB-31
Animal Production (includes Fur-Bearing Animals)	AB-32
Bacteriology, General and Systematic	AB-45
Behavioral Biology	AB-46
Biochemistry	AB-54
Biophysics	AB-79
Blood, Blood-Forming Organs and Body Fluids	AB-87
Bones, Joints, Fasciae, Connective and Adipose Tissue	AB-104
Botany, General and Systematic	AB-122
Cardiovascular System	AB-133
Chemotherapy	AB-182
Chordata, General and Systematic Zoology	AB-199
Circadian Rhythm and Other Periodic Cycles	AB-209
Cytology and Cytochemistry	AB-209
Dental and Oral Biology	AB-219
Developmental Biology-Embryology	AB-224
Digestive System	AB-236
Disinfection, Disinfectants and Sterilization	AB-260
Ecology (Environmental Biology)	AB-262
Economic Botany	AB-296
Economic Entomology (includes Chelicerata)	AB-297
Endocrine System	AB-308
Enzymes	AB-336
Evolution	AB-347
Food and Industrial Microbiology	AB-348
Food Technology (non-toxic studies)	AB-359
Forestry and Forest Products	AB-373
General Biology	AB-379
Genetics of Bacteria and Viruses	AB-380
→ Genetics and Cytogenetics	AB-400
Gerontology	AB-453
Horticulture	AB-453
Immunology (Immunochemistry)	AB-466
Immunology, Parasitological	AB-521
Integumentary System	AB-526
Invertebrata, Comparative and Experimental Studies	AB-532
Invertebrata, General and Systematic Zoology	AB-554
Laboratory Animals	AB-575
Mathematical Biology and Statistical Methods	AB-575
Medical and Clinical Microbiology (includes Veterinary)	AB-575
Metabolism	AB-604
Methods, Materials and Apparatus, General	AB-612
Microbiological Apparatus, Methods and Media	AB-613
Microorganisms, General (includes Protista)	AB-614*
Morphology and Anatomy of Plants (includes Embryology)	AB-614
Morphology and Cytology of Bacteria	AB-615
Muscle	AB-615
Neoplasms and Neoplastic Agents	AB-625
Nervous System (excludes Sense Organs)	AB-707
Nutrition	AB-791
Paleobiology	AB-801
Paleobotany	AB-802
Paleozoology	AB-803
Palynology	AB-805
Parasitology (includes Ecto- and Endoparasites)	AB-805
Pathology, General and Miscellaneous	AB-811
Pediatrics	AB-812
Pest Control, General (includes Plants and Animals); Pesticides; Herbicides	AB-813
Pharmacognosy and Pharmaceutical Botany	AB-815
Pharmacology	AB-818
Physical Anthropology: Ethnobiology	AB-906
Physiology and Biochemistry of Bacteria	AB-906
Physiology, General and Miscellaneous	AB-919
Phytopathology	AB-922
Plant Physiology, Biochemistry and Biophysics	AB-944
Poultry Production	AB-969
Psychiatry	AB-972
Public Health	AB-995
Radiation Biology	AB-1049
Reproductive System	AB-1057
Respiratory System	AB-1075
Sense Organs, Associated Structures and Functions	AB-1094
Social Biology (includes Human Ecology)	AB-1118
Soil Microbiology	AB-1119
Soil Science	AB-1122
Temperature: Its Measurement, Effects and Regulation	AB-1132
Tissue Culture: Apparatus, Methods and Media	AB-1134
Toxicology	AB-1134
Urinary System and External Secretions	AB-1172
Veterinary Science	AB-1184
Virology, General	AB-1185

*There are no references listed under this heading. See page listed for instructions on locating related topics.

FIGURE 13.

Reproduction of a page from the beginning of *Biological Abstracts* showing organisation of abstracts. The arrow indicates the topic containing the reference by Lesueur *et al.* (1991).

Unfortunately, an appropriate key word doesn't always lead to a useful reference. To the right and left of the key word *prolactin*, for example, and for the 50 lines under this key word, you will find a few words or parts of words (drawn from each paper's title with terms added by BIOSIS editors) that partially inform us of each paper's content. Each of these papers has something to do with prolactin and the trick is to figure out what.

The only way to tell whether or not a reference is relevant to our topic is to look it up. Such listings provide many ambiguous leads, each of which must be investigated. One entry, indicated in Figure 10 by an arrow, does appear somewhat promising: RECEPTOR ON PROLACTIN-IND . . . AND SHORT FORMS OF THE

Although the paper's precise subject is a bit of a mystery, it does seem to be about prolactin receptors. We turn to reference no. 95343 as directed and are rewarded with the listing indicated with an arrow in Figure 11 (page 46). (We could also have arrived at this listing by searching under the keyword 'milk', as indicated by the arrows in Figure 12, page 47.) Not only does *Biological Abstracts* provide us with the complete citation for the paper by Lesueur *et al.*, but it gives us an informative abstract as well, summarising the contents of the paper (Figure 11). These abstracts are grouped by subject, as indicated in Figure 13 (previous page); we might, therefore, also have discovered the Lesueur *et al.* paper by simply scanning all the abstracts grouped under the heading 'Genetics and Cytogenetics'.

Note that *Biological Abstracts* can be used to advantage in combination with other indexing services. If you locate an appropriate reference in *Science Citation Index,* for example, you can retrieve an abstract of that paper in *Biological Abstracts. Biological Abstracts* is currently published in 24 bi-monthly issues compiled in two volumes per year; each is about two inches thick – which gives you some idea of the tremendous rate at which research articles are now being published. It is also available on CD-ROM.

Using *Current Contents*

One additional search tool of value particularly to final year undergraduates and to post-graduate biology students is *Current Contents*, published by the Institute for Scientific Information in Philadelphia, PA. *Current Contents* essentially puts the best library in the world at your fingertips. Each weekly issue includes the complete table of contents for scientific journals published a few weeks earlier; over 900 different journals are covered by the publication, so you are not likely to miss much of the relevant literature in your field, no matter how meagre the holdings in your institution's library. If you encounter a paper of particular interest while browsing through the latest issue of *Current Contents,* you can try to find that article in your library or, failing that, you can request a copy of the paper through the inter-library loan service or from its author (mailing addresses are given at the back of each issue). Also towards the back of each issue is a subject index, allowing you to locate papers concerning particular topics very quickly. A computerised version of *Current Contents* has been introduced on CD-ROM and in the BIDS service.

CLOSING THOUGHTS

Printed indexing services, *Current Contents,* and computerised data bases are all good sources of references, but there is a major catch: your library will probably not subscribe to all – or even most – of the journals included in the literature searched by the various services. Although *Current Contents* provides a partial solution to this problem for very recent literature, you must be able to wait from one to several weeks or more before having the actual research paper in hand. You may spend considerable time accumulating a long list of intriguing references, only to discover that most of them are not to be found on your campus or in any other nearby library. Consulting recent issues of available, appropriate journals may thus be the most efficient way to search for promising research topics and references.

SUMMARY

1. Become a 'brain-on' reader: work to understand your sources fully, sentence by sentence, figure by figure, and table by table.

2. Try to take notes thoughtfully, in your own words.
3. In your note-taking, be careful to distinguish your words and thoughts from those of the author(s).
4. Be efficient in exploring the primary scientific literature: browse the list of references given in your textbook, other relevant books, and the papers published in recent issues of relevant scientific journals.
5. Become familiar with the major abstracting and indexing services, including computerised data bases, and use these as necessary to complete your literature search.

3. Writing Up Your Experiments

CLASS LABORATORY REPORTS, LARGER PROJECT REPORTS AND HIGHER DEGREE THESES

Class lab reports are usually based on one or a few related experiments. You will usually be given a schedule (protocol, handout or lab manual) with a title, introduction and methods. Some schedules have results tables for you to fill in, and a few elementary ones require you to do nothing more than fill in the gaps on the printed sheets. For other experiments, you may need to write everything, including the title, introduction, methods, results, analysis and conclusions. Even if an introduction and a methods section are included in the schedule, you may wish to add details of your own, and must do so if the methods actually used differed from those in the schedule. A Literature Cited section is not needed for most basic undergraduate reports, but is for more advanced ones. For lab reports, take advice from the staff on the format, approximate length and style.

You should write up all experiments and problem sheets, for your own benefit, whether they count for course assessment marks or not, because writing them makes you concentrate on what you did, why and how. It also reinforces the theory part of the course. Write these reports as soon as possible after completion of the experiments, while the details are fresh in your mind. Particularly in your first undergraduate year, get as much feed-back as you can on any reports you hand in, to improve your style before you have to write important final-year research project reports. In such larger project reports, covering several weeks or more of research on a specific project, a more formal style is required, with a greater need to cite references for facts, and better organisation to cope with the larger amount of information. You will then have to write your own title, introduction (including a literature survey), methods and materials, analysis, conclusions and literature cited sections.

MSc, MPhil and PhD theses must be written in a formal style, with rigorous referencing of facts, and in the style prescribed by the regulations of your faculty of your university. Consult the university regulations – the library will have a copy – on whether there is a maximum number of words for a thesis, and on the page size, binding, etc. Consult a library copy of a recent thesis as a general guide on the format and style. For a thesis, your records of what you have done, how and why, need to be detailed, clear and accessible. It is very easy to forget details of an experiment performed early in your project when you come to write the final thesis, so keeping a good lab notebook is essential.

THE VALUE OF WRITING LAB REPORTS

Doing biology involves asking questions, formulating hypotheses, devising experiments to test the hypotheses, and presenting, evaluating and interpreting data. Those so-called facts you learn from lectures and textbooks are primarily interpretations of data. By participating in the acquisition and interpretation of data, you glimpse the true nature of the scientific process.

If you are contemplating a career in research, be assured that learning to write effective lab reports now is an investment in your future. Later on, perhaps as a lab technician or research assistant, you will often be asked to present data so that the future path of the research can be decided upon. Even a PhD thesis, unless purely about theory, is essentially a large lab report. Writing up your research for publication, as a post-graduate student or as a professional biologist, you will quickly find that you are following the procedures you used in preparing good undergraduate lab reports.

Writing lab reports develops your ability to organise ideas logically, to think clearly, and to express yourself accurately and concisely. It is difficult to imagine a career in which mastery of such skills is not a great asset.

ORGANISING A LABORATORY OR FIELD NOTEBOOK

The first step in preparing a good lab report is to keep a detailed notebook. The functions of this notebook are: to record the design and goals of your experiments; to record and organise your thoughts and questions about the work; to help you organise your activities in the lab so that you can work quickly and accurately; to record your observations, drawings and numerical data, and your analyses and conclusions.

Keeping a detailed notebook will make the writing of your report much easier and you will produce a better report. The skills you learn in keeping the notebook could really pay off in later life. The Nobel Prize for the isolation of insulin would probably have gone to J. B. Collip instead of F. Banting and C. H. Best, had Dr. Collip learned as a student to keep a more careful record of his work. Collip was apparently the first to purify the hormone, but his notes were incomplete and he was unable to repeat the procedure successfully in subsequent studies. Don't let something like that happen to you!

In keeping your notebook, assume that you are doing something worthwhile, that you might discover something remarkable, and that you will suffer complete amnesia that night. In other words, take the time to write down – in your own words – everything you are about to do, everything you actually do and why you do it. Record your data clearly, with each number identified by the appropriate units. Many of the details that seem too obvious to write down (*i.e.* the name of the species you are working with, or the unit of measurements – centimetres rather than inches, for example) are forgotten surprisingly quickly upon leaving the lab. You should normally work in SI units – see page 55.

Write so legibly and clearly that should you be run over the next day, other students on the course would have no difficulty reconstructing your study and following your results. If you use abbreviations, be sure to indicate what each stands for. These procedures greatly facilitate the writing of your report and are crucial in any research lab, since others must be able to pick up your work or interpret it where you left off if you are unable to come in or if you leave that lab.

Some of this writing can be done before the lab session. Whenever possible, read about the day's study ahead of time, being sure (through writing in your notebook) that you understand its goals, and plan exactly how you will record your data. You will get more out of the exercise and will finish your work sooner if you arrive prepared. It often helps to sketch a simple flowchart of your planned activities ahead of time, as in Figure 14 (next page).

Your notebook should contain any thoughts, observations, or questions you have about what you are doing, along with the actual schedule and data. A sample page from a model notebook is shown in Figure 15 (page 57). With such clear, well-organised, and well thought-out notes, this student is well on the way to preparing a fine lab report.

A notebook recording field observations (a 'field notebook') would look similar to that shown in Figure 16 (pages 58–9), which records part of a study investigating the size distributions of the common marine intertidal snail, *Littorina littorea,* along a Massachusetts beach. Again, the quality of the entries suggests that this student is not simply trying to 'get it over with', or to write down textbook facts and paraphrases of lecture notes. The student is clearly looking around and thinking – every lecturer's dream!

SI UNITS AND PREFIXES

Unless instructed otherwise, you should always use SI units (Système International d'Unités). There are seven basic units, given here with their abbreviations: length, metre, m; mass, kilogram, kg; time, second, s; temperature interval, kelvin, K; electric current, ampere, A; amount of substance, mole, mol; luminous intensity, candela, cd. There are also derived units, from combining two or more basic units, such as speed, which is measured in metres per second. Some derived units are given their own names and symbols for convenience: power, for example, is expressed in watts, symbol W, derived from one kilogram multiplied by one metre

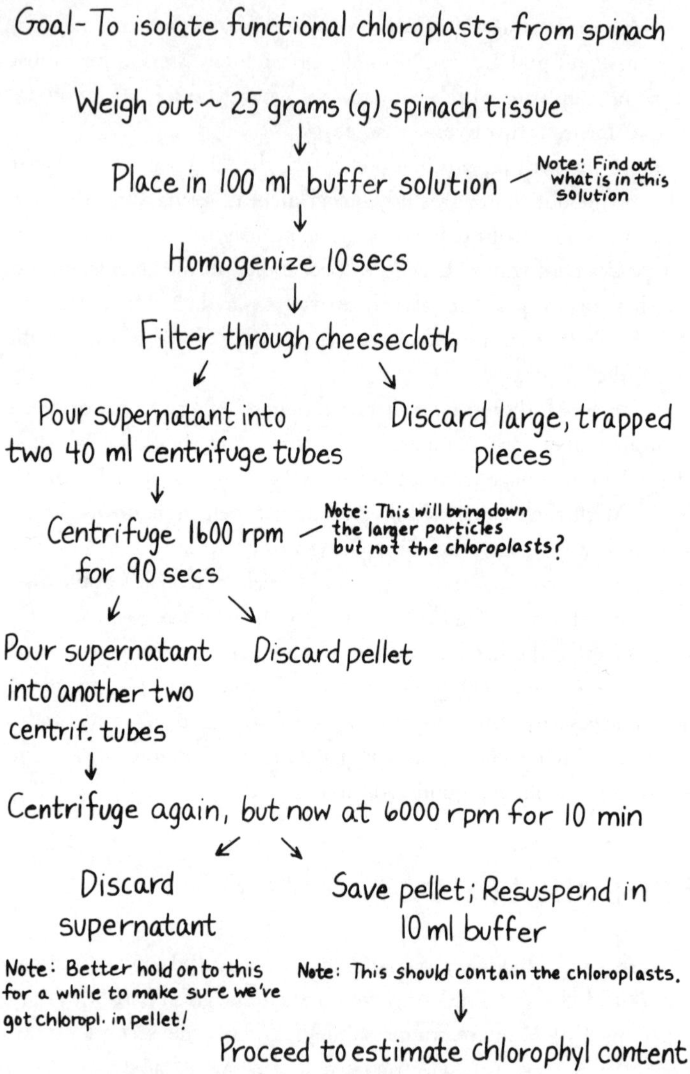

Goal- To isolate functional chloroplasts from spinach

Weigh out ~ 25 grams (g) spinach tissue

↓

Place in 100 ml buffer solution — Note: Find out what is in this solution

↓

Homogenize 10 secs

↓

Filter through cheesecloth

↙ ↘

Pour supernatant into two 40 ml centrifuge tubes | Discard large, trapped pieces

↓

Centrifuge 1600 rpm for 90 secs — Note: This will bring down the larger particles but not the chloroplasts?

↙ ↘

Pour supernatant into another two centrif. tubes | Discard pellet

↓

Centrifuge again, but now at 6000 rpm for 10 min

↙ ↘

Discard supernatant

Note: Better hold on to this for a while to make sure we've got chloropl. in pellet!

Save pellet; Resuspend in 10 ml buffer

Note: This should contain the chloroplasts.

↓

Proceed to estimate chlorophyl content

FIGURE 14.

A page from a student's lab notebook. The flowchart is based on more detailed information provided by the lecturer.

October 22, 1991

Goal -- to determine rate at which the marine bivalve Mercenaria mercenaria (hardshell clam) moves water across its gills.

Approach -- Use unicellular alga (phytoplankton) Dunaliella tertiolecta. Determine initial cells · ml^{-1}, final cells · ml^{-1}. If know elapsed time and volume of seawater in container, can calculate cells eaten per hour per clam, and ml of sea H_2O cleared of cells/h/clam.

Weight of clam (incl. shell): 9.4 g [Fisher/Ainsworth balance, Model MX-200]

Find initial concentration of algal cells:

1. 1 ml of culture + 1 drop of Lugol's iodine to kill cells.
2. Load hemacytometer for cell counts -- finally got it right on 5th try! Helps to tilt pipet at about 45° angle.

Data: (multiply by 1.05 [because of dilution w/0.05 ml Lugol's sol'n] and then by 10^4 ---→ cells · ml^{-1})

Note: ask Prof. Scully why 10^4.

counts per section of hemacytom slide

	Sample 1	Sample 2
	22	18
2)	13	21
	18	20
4)	22	20
	26	16
6)	21	12
	20	18
8)	16	22
	22	19
$\bar{X}$ =	20.0 →	18.4 →
→	21.0×10^4 cells · ml^{-1}	19.4×10^4 cells · ml^{-1}

good agreement!

Put clam in 150 ml of this solution at 2:10 PM; H_2O = 21.4°C

FIGURE 15.

A sample page from a student's lab notebook.

September 17, 1991. 1–3 PM. Sunny, 27°C air temp.; 17°C water.
Goal: investigate size distribution of _L. littorea_ shells at different distances from high tide line.
ME: might expect shells to get larger as get closer to water, since these are marine snails and so should be able to feed more hrs per day if spend more time under water. (What do they eat? Can they only eat in water?)

Note: Many of the empty shells have large round holes made by a drilling predator -- probably _Nucella lapillus_ or _Lunatia heros_, which seem to be the only carnivorous gastropods living here.
⟶ Should calc. % of dead snails (empty shells) that have drill holes. Drill holes seem to be in exact same location on each shell. How is this possible?

Note: Most of the rocks in this area are almost completely covered by very tiny (young) barnacles. Covering = so dense in places can't even see the rock surface.
Question: Why no big, adult barnacles on these rocks? Perhaps all die off during the summer for some reason. There _are_ many large barnacles on rocks much further down the beach. Same species? If so, why only die here? Would be interesting to come back and monitor survival of young barnacles here and down the beach a few times during the summer and fall -- perhaps once a month?

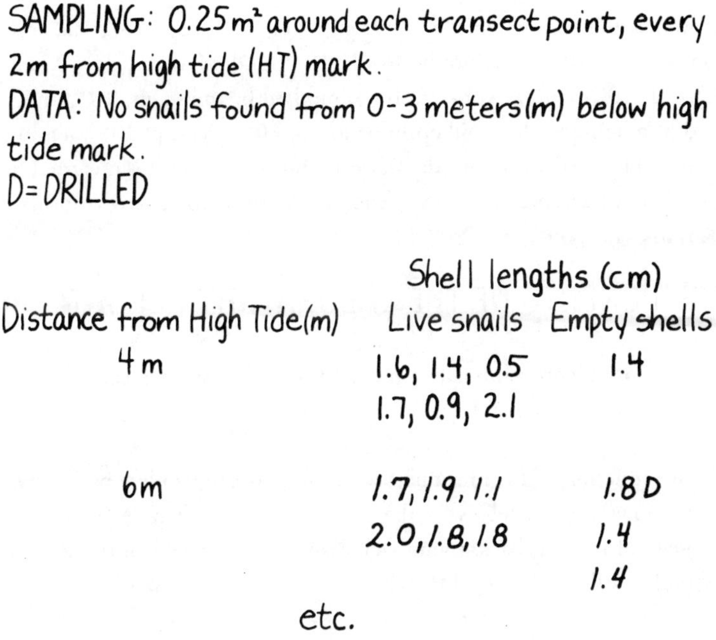

SAMPLING: 0.25 m^2 around each transect point, every 2m from high tide (HT) mark.
DATA: No snails found from 0-3 meters (m) below high tide mark.
D = DRILLED

	Shell lengths (cm)	
Distance from High Tide (m)	Live snails	Empty shells
4m	1.6, 1.4, 0.5 1.7, 0.9, 2.1	1.4
6m	1.7, 1.9, 1.1 2.0, 1.8, 1.8	1.8 D 1.4 1.4

etc.

FIGURE 16.

(Left and above) A sample page from a student's field notebook.

squared divided by one second cubed. Consult a good dictionary, such as Chambers 20th Century Dictionary, or an encyclopaedia, or other reference book for more details. Although concentrations should be expressed in moles (molarity is the number of moles of dissolved substance per litre of solution) or millimoles, some staff prefer to use grams per litre, as that is easier to use when weighing out compounds for making up solutions.

Abbreviations of compound units are written with a small space between the units, *e.g.* N m for a newton metre. For units to be divided, negative indices are preferred, but a solidus, or 'slash' (/) may be used, *e.g.* a newton (unit of force) can be defined as one kg m s^{-2} or as one kg m/s^2. Multiples and submultiples are indicated by standard prefixes, the more common of which are given here with their abbreviation and multiplication factor: tera-, T, 10^{12}; giga-, G, 10^9; mega-, M, 10^6, million; kilo-, k, 10^3, thousand; hecto-, h, 10^2, hundred; deca-, da, 10, ten; deci-, d, 10^{-1}, tenth; centi-, c, 10^{-2}, hundredth; milli-, m, 10^{-3}, thousandth; micro-, μ, 10^{-6}, millionth; nano-, n, 10^{-9}; pico-, p, 10^{-12}; femto-, f, 10^{-15}. The prefixes hecto-, deca-, deci- and centi – are not preferred SI units, but are accepted for convenience, as in centimetre and decinormal.

COMPONENTS OF THE LABORATORY REPORT

A laboratory report, especially a more advanced one, is typically divided into five major sections:

1. **Introduction.** The introductory section, usually only one or two paragraphs long, tells why the study was undertaken. A brief summary of relevant background facts leads to a statement of the specific problem that is being addressed.

2. **Materials and methods.** This section is *your* reminder of what you did, and it also serves as a set of instructions for anyone wishing to repeat your study. It may include drawings of the equipment or of the experimental set-up.

3. **Results.** This is the centre-piece of your report. What were the major findings of the study? Present the data or summarise your observations, using graphs and tables to reveal any trends you found. Point out these trends to the reader. If you make good use of drawings, tables and graphs, the results can usually be presented in only one or two paragraphs of text. One picture can often replace many words. Avoid interpreting the data in this section.

4. **Discussion.** How do your results relate to the goals of the study, as stated in your introduction, and how do they relate to the results that might have been expected from background information obtained in lectures, textbooks or outside reading? What new hypotheses might now be formulated, and how might these hypotheses be tested?

5. **Literature Cited.** This section includes the full citations for any references (including textbooks and lab handouts) you have mentioned in your report. Check your sources to be certain they are listed correctly, because this list will permit the interested reader to check the accuracy of any factual statements you make. Ideally, cite only material you have actually read.

In a thesis, but not normally in a class lab report, two additional sections are included: the Abstract, summarising the nature of the problem addressed, your approach to the problem, and the major findings and conclusions; and an Acknowledgments section, in which you formally thank people for their contributions to the project.

Before writing your first report, it is helpful to study a few short papers in a major biological journal, such as *Developmental Biology*, *Ecology*, or *Genetics*. Reading the journal articles for content is unnecessary; you don't need to understand the topic of a paper to appreciate how the article is crafted. But do pay attention to the way the Introduction is constructed, the amount of detail included in the Materials and Methods section, and the material that is and is not included in the Results section. Note that figures and tables are always accompanied by explanatory captions, and that the axes of graphs and the columns and rows of tables are clearly labelled.

WHERE TO START

Strangely enough, the Introduction is not the place to begin writing your report; it is far easier to write the Introduction towards the end, after you have digested what it is that you have done. Start work on the Materials and Methods section or on Results, or on the two in tandem. Working on the Results section may help to clarify what should be included in the Methods section, and working on the Methods section may clarify the order in which results should be presented.

Because the Materials and Methods section requires the least mental effort, completing it is a good way to overcome inertia. Even if you do not know why you did the experiment or what you found out by doing the experiment, you can probably reconstruct what you did without much difficulty, from the schedule and your notebook!

THE MATERIALS AND METHODS SECTION

Results are meaningful in science only if they can be obtained over and over again, whenever the experiment is repeated. Unfortunately, the results of any study depend to a large extent on the way the study was done. It is therefore essential that you describe your methodology in detail sufficient to permit your experiment to be repeated exactly. Writing a detailed Materials and Methods section helps you to review what you have done in an organised way and starts you thinking about why you've done it. The difficulty in writing this section is in selecting the right level of detail. Students commonly give too little information; when informed of this defect, they may then give too much information. Keeping your audience in mind (yourself and your fellow students) will help.

Determining the Correct Level of Detail

Many students begin with a one-sentence Materials and Methods section: 'Methods were as described in the lab schedule.' Although this is brief, it is often unacceptable. Studies are rarely performed exactly as described in a lab manual or handout. Your instructions may call for the use of 15 animals, but only 12 might be available for use. Many details will vary from year to year, week to week, or place to place, and must therefore be given by you in your report. But don't get carried away! Consider the following over-detailed description of a study involving the growth of radish seedlings.

> On January 5, I obtained four paper cups, 400 g of potting soil, and 12 radish seeds. I labelled the cups A, B, C, D, and planted three seeds per cup, using a plastic spoon to cover each seed with about one-quarter inch of soil.

The author has used the first sentence simply to list the materials;

whenever possible, it is far better to mention each new material as you discuss what you did with it. Also, why do we need to know the weight of the soil obtained, or that the cups were labelled A – D rather than 1 – 4, or that a plastic spoon was used to add soil? Omitting the excess details and starting with what was done, we get:

> On January 5, I planted three radish seeds in each of four individually-marked paper cups, covering the seeds with about one-quarter inch of potting soil.

Note that the essential details – individually-marked cups, three seeds per cup, one-quarter inch of soil – not only survive in the edited version, but stand out clearly. The skill is to determine which details are essential and which are not.

The best approach to writing the Materials and Methods section is to list all of the factors that might have influenced your results. If, for example, you measured the feeding rates of caterpillars on several different diets, your list might look something like this:

- Species of caterpillar used
- Diets used
- Amount of food provided per caterpillar
- Time of year
- Time of day
- Air temperature in room
- Manufacturer and model number of any specialised equipment used (such as balances, centrifuges, or spectrophotometers)
- Size and age of caterpillars
- Duration of experiment
- Container size
- Number of animals per container
- Total number of individuals in study

This list, which you do not hand in with your report, contains the bricks with which you will construct the Materials and Methods section. Each of the listed details must find its way into your report (not necessarily in the order in which you jotted them down) because each gives information essential for later replication of the experiment. Some of this information may also help you explain in the Discussion section why your results differ from those of others who have gone before you, if they do differ.

In describing the procedures followed, you must say what you did, but you should freely refer to your lab manual or handouts in describing how you did it. For example, you might write:

> The three different diets were distributed to the caterpillars in random fashion, as described in the lab manual (Roberts, 1992).

The important point here is that the diets were distributed at random; the outcome might be quite different if the largest caterpillars were to receive one diet and the smallest caterpillars another. The interested reader can refer to the stated source (Roberts, 1992) for detailed instruction in the method of randomisation. You should routinely include the relevant portion of your handout, schedule or manual with your report.

It is often a good idea to mention, for your own benefit as well as that of your reader, why particular steps were taken. Imagine yourself explaining things to a classmate. We might, for example, rewrite the example in the preceding paragraph to read:

> To avoid prejudicing the results by distributing food according to size of caterpillar, the three different diets were distributed to the caterpillars in random fashion as described by Roberts(1992).

It is also usually appropriate to include any formulae used in analysing your data. The following sentences, for example, would belong in a Materials and Methods section.

> The data were analysed by a series of Chi-Square tests. The rate at which food was eaten was calculated by dividing the weight loss of the food by the time available (three

hours), according to the following formula:

Feeding rate = (Initial food weight – final food weight) ÷ 3 h.

Be sure to note any departures from the given instructions. Suppose you were told to weigh the caterpillars individually but found that your balance was not sensitive enough to record a single animal. Your lecturer would probably suggest weighing the individuals in each container as a group. Your report might then include the following information:

> Determining the weight gained by each caterpillar over the three-hour period of the experiment required that both initial and final weights be determined. The caterpillars were too small to be weighed individually. Therefore, similarly sized caterpillars were weighed in groups of three at a time. The average weight of each caterpillar in the group was then calculated.

A Model Materials and Methods Section

The Materials and Methods section should be brief but informative. The following example completely describes an experiment designed to test the influence of decreased salinity on the body weight of a marine worm.

> The polychaete worms used in this study were Nereis virens, freshly collected from Nahant, MA, and ranging in length between 10 and 12 cm. All treatments were performed at room temperature, approximately 21°C, on April 22, 1993. One hundred ml of full-strength sea-water was added to each of three 200 ml glass jars, which served as controls, to monitor worm weight in the absence of salinity change. Another three jars were filled with 100 ml of sea-water diluted by 50% with distilled water.
>
> Six polychaetes were quickly blotted with paper towels to remove adhering water, and were then weighed to the nearest 0.1 g using a Model MX-200 Fisher/Ainsworth balance. Each worm was then added to one of the jars of

> sea-water. Blotted worm weights were later determined at 30, 60, and 120 minutes after the initial weights were taken.
>
> The initial and final osmotic concentrations of all test solutions were determined using a Model 3W Advanced osmometer, following instructions provided in the handout (Lima, 1992).

All essential details have been included: temperature, species used, size of animals used, number of animals per treatment and per container, volume of fluid in the containers, type and size of containers, time of year, and equipment used. After reading this Materials and Methods section, you could repeat the study if necessary. The writer has made clear why certain steps were taken; three jars of full-strength sea-water served as controls, for example, and worms were blotted dry to remove external water. The fact that worms were blotted dry before they were weighed was mentioned because this is a procedural detail that would obviously influence the results. On the other hand, the author does not describe how the balance was operated, since this technique is standard. The author has written a report that might be useful to him or her in the future . . . and gets a top grade.

In some disciplines it is usual to list the strains and sources of the organisms, such as different inbred lines of mice, as these differ in various characteristics, or to list the catalogue or isolation number of a particular mutant, such as mutation *asco* (37402) in the fungus *Neurospora crassa*, where the isolation number 37402 distinguishes this one mutation from others in the same gene.

THE RESULTS SECTION

Here you summarise your findings, using drawings, tables, graphs and words. The Results section is not the place to discuss why or how the experiment was performed, nor whether the results were expected, unexpected, disappointing, or interesting. Simply present the results, drawing the reader's attention to the major observations and key trends in the data. Don't interpret them here.

Summarising Data Using Tables and Graphs

Before you write your Results section, you must work with your data, which probably contain a story that is crying out for recognition. The purpose of tabling and graphing data is not to add bulk to lab reports, but is to manipulate the data in order to reveal the underlying trends and story.

There is no single right way to present summaries of data; use whatever system gives the clearest illustration of trends. You must decide what relationships might be worth examining and then experiment with different ways of tabulating and graphing the data to best explore and demonstrate those relationships. Suppose we return to the experiment in which we measured the rates at which caterpillars fed for three hours on three different diets. We determined both the initial weight of food provided and the weight of food remaining after the three-hour period, so we can calculate the weight of food eaten per caterpillar per hour. In your report, you should provide a sample calculation, so that if you make a mistake your lecturer can see where you went wrong. We know the initial and final weights of the caterpillars for each diet, and the initial and final weights of dishes of food in the absence of caterpillars; these control dishes will tell us the amount of water lost by the food because of evaporation.

What relationships in the data might be especially worth examining? The first step in answering this question is to make a list of specific questions that might be worth asking:

1. Do the caterpillars feed at different rates on the different diets? That is, does feeding rate vary with diet?
2. Do larger caterpillars eat faster than smaller caterpillars? That is, does feeding rate vary with size of caterpillar?
3. How is the weight gained by a caterpillar related to the weight lost by the food?
4. Did the weight of the control dishes change; if so, by how much?

As in preparing the Materials and Methods section, this list is for your own use and is not included in your report. Write these questions in complete sentences. Once you have this list, it is easy to list the relationships that must be examined:

Table 1. Summary of Raw Data

Diet	Initial Caterpillar Wt. (g)	Final Caterpillar Wt. (g)	Caterpillar Weight Change (g)	Wt. of Food Lost (g) over 3 h	Feeding Rate (g food lost/h caterpillar)
A	8.05	9.55	+1.55	3.65	15.2×10^{-2}
A	4.80	5.80	+1.00	1.74	7.2×10^{-2}
A	5.50	7.00	+1.50	3.33	13.9×10^{-2}
A	5.50	4.70	~~−0.80~~	~~0.00~~	0
A	5.90	6.95	+1.05	1.35	5.6×10^{-2}
Average	5.95	6.80	+1.28	2.52	8.4×10^{-2}
B	4.40	5.11	+0.71	2.19	9.1×10^{-2}
B	5.20	5.60	+0.40	1.25	6.2×10^{-2}
⋮	⋮	⋮	⋮	⋮	⋮
Control 1	—	—	—	0.22	—
2	—	—	—	0.10	—
3	—	—	—	0.16	—

1. Feeding rate as a function of diet
2. Feeding rate as a function of caterpillar size
3. Caterpillar weight gain versus food weight loss for each caterpillar
4. Food weight loss in the presence of caterpillars versus food weight loss in controls

Constructing a Summary Table

Now you must organise your data into a table in a way that will let you examine each of these relationships. Consider Table 1. This rough draft lists all the data obtained in the experiment. For the first relationship in our list (feeding rate as a function of diet), a table will tell the entire story. You can simply present a summary table, with explanatory caption, as in Table 2. Note that one of the caterpillars offered diet *A* ate no food and lost weight during the experiment. This individual died during the same study and the associated data were therefore omitted from Table 2. (The weight loss for this caterpillar probably reflects evaporation of body water.)

Table 2. Average rates of food consumption over a 24 h period for caterpillars given three separate diets.

Diet	No. Caterpillars	(g food eaten/caterpillar/h)
A	4*	8.4×10^{-2}
B	5	3.8×10^{-2}
C	5	7.9×10^{-2}

*One individual died during the study, without eating any food.

To Graph or Not to Graph

Finally, the time has come to reveal more subtle trends, which may not be readily apparent from the summary table (Table 2), but which may be made visible through graphing. But do not automatically assume that your data must be graphed. If you can tell your story clearly using only a table, a graph is superfluous. You may be able to summarise some aspects of the data without graphs or tables. You might write, for example, 'No animals ate at temperatures below 15°C', and present data only for animals at higher temperatures.

Graphs in biology generally take one of two basic forms: scatter plots (point graphs) or histograms and bar graphs. For the second relationship (feeding rate versus caterpillar size) we wish to examine in the caterpillar data, a scatter graph, like Figure 17, will be especially appropriate.

Note that each axis of the graph in Figure 17 is clearly labelled, including units of measurement; the meaning of each symbol is clearly

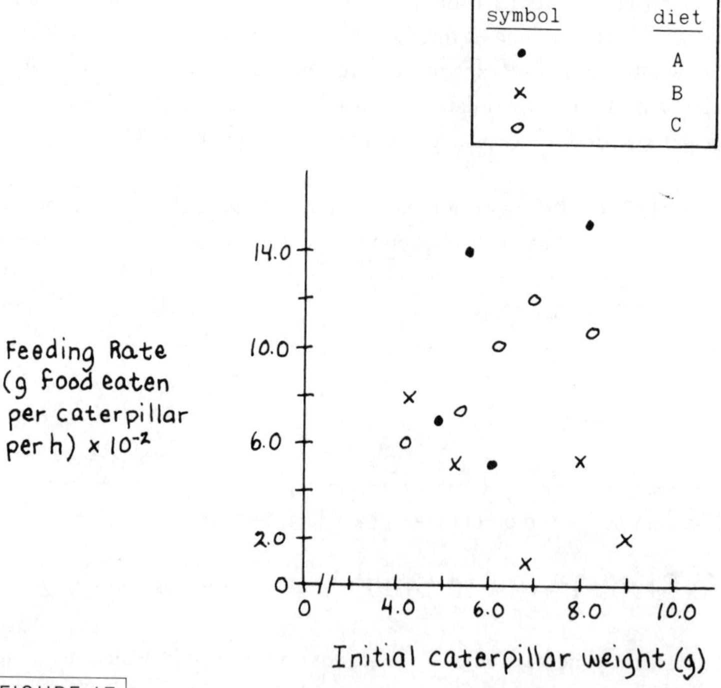

FIGURE 17.

The relationship between initial caterpillar weight and rates of food consumption for *Manduca sexta* feeding on three different diets at 24°C.

indicated in the key, and a detailed explanatory caption ('figure legend') accompanies the figure. In Figure 17, it would be insufficient to label the *Y*-axis 'Feeding Rate'. Feeding rates can be expressed as per minute, per hour, per day, or per year, and can be expressed as per animal, per group of animals, or per gram of body weight. Similarly, it is unacceptable to label the *X*-axis as 'Weight', or even as 'Caterpillar Weight'. From the figure caption, the axis labels, and the graph itself, the reader should be able to determine the question being asked, get a good idea of how the study was done, and be able to interpret the figure without reference to the text. A good graph is self-contained.

The third relationship (animal weight gain versus food weight loss) might well be left in table form, since the trend is readily discernible; caterpillars always gained less weight than that lost by the food. The same

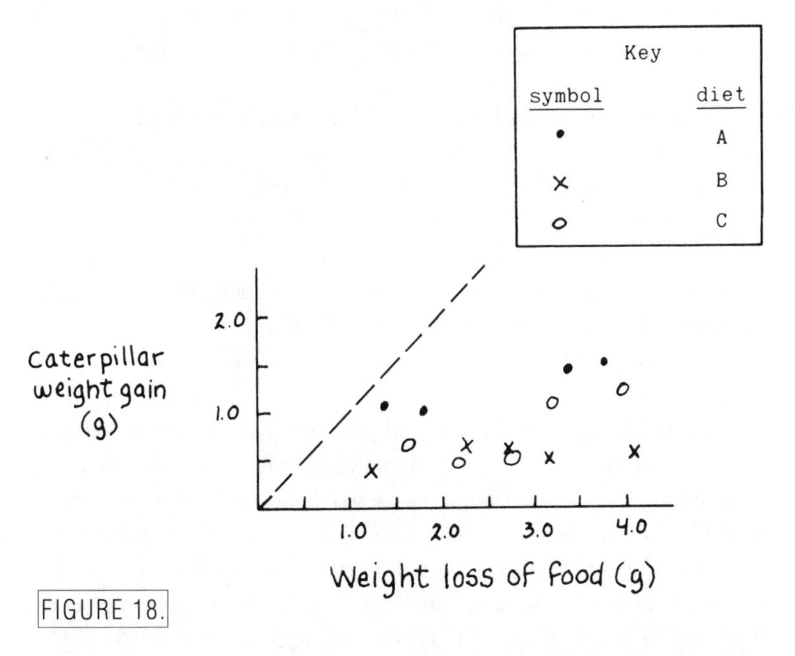

FIGURE 18.

Caterpillar weight gain as a function of food consumption for *Manduca sexta* fed on any one of three diets at 24°C. Points falling on the dotted line would indicate equality between weight gain and food eaten.

trend could be revealed more dramatically (more graphically) with a scatter plot, as shown in Figure 18 (previous page), but a graph is not essential here. Again, note the steps taken to avoid ambiguity: the axes are labelled, units of measurement are indicated, symbols are interpreted on the graph, and the figure is accompanied by an explanatory legend. The symbols in Figure 18 are consistent with their usage in Figure 17. Always use the same system of symbols throughout a report, so as not to confuse your reader; if filled circles are used to represent data obtained on diet *A* in one graph, filled circles should be used to represent data obtained on diet *A* in all other graphs.

The fourth relationship in our list concerns food weight loss in control dishes (no caterpillars). No graphs or tables are needed here; two sentences will do:

> Control containers exhibited less than a 3% weight loss (N = 3 containers) during the 24h period. In contrast, food in containers with caterpillars lost at least 23% of initial weight.

If the weight loss had been substantial, perhaps 5–10 per cent or more of initial weight, you might wish to adjust all the data in your tables accordingly, before making other calculations:

> Control containers exhibited a 7.6% weight loss (N = 3 containers) over the 24h period. We therefore adjusted weight loss in other containers for this 7.6% evaporative loss before calculating feeding rates.

You would then provide a sample calculation so that your lecturer could see how this was done, and so that you can be sure of what you did if consulting your report later. A less desirable but acceptable alternative would be to state the magnitude of the evaporative weight loss in your Results section and bring this point up again in interpreting your results in the Discussion. In this case, you might want to label appropriate portions of graphs and tables as 'Apparent Feeding Rates' rather than 'Feeding Rates'. There may be several 'right ways' to present the data; you must simply be complete, logical and clear.

So far, we have looked only at examples of tables and point plots. If

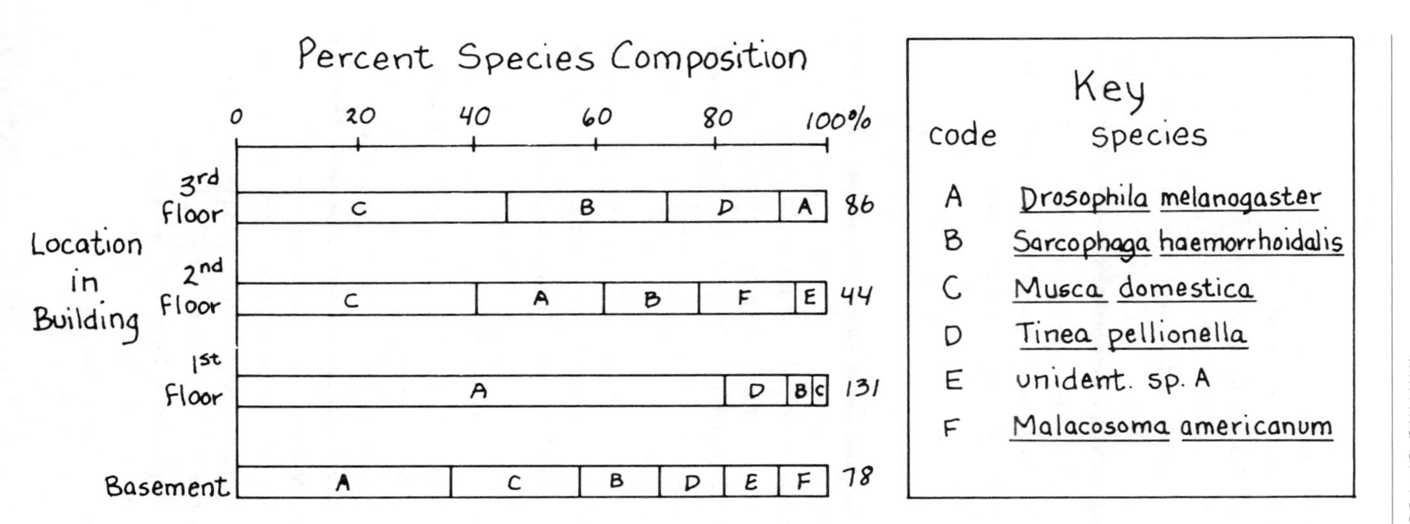

FIGURE 19.

The distribution of insect species collected from light fixtures on four floors of the Biology building. The number to the right of each bar gives the total number of insects collected on each floor.

you were studying the differences in species composition of insect populations trapped in the light fixtures on four different floors of your Biology Department building, a bar graph, as in Figure 19 (previous page), might be more suitable. The axes are clearly labelled, including units of measurement, and an explanatory legend accompanies the figure. The graph tells an interesting story: because *A* is the fruit fly *Drosophila melanogaster*, it is not difficult to deduce where the genetics lab is located! Use tables and graphs only if they make your data work for you. Don't include a drawing, graph or table unless you plan to discuss it. Include only those illustrations that best help you to tell your story.

Preparing Graphs

Graphs may be constructed with the aid of a computer, but unless your lecturer suggests otherwise don't feel that you *must* submit computer-generated graphs to earn a top mark. Most staff would rather see a carefully thought-out and neatly executed graph done by hand than a poorly thought-out piece of computer graphics. To emphasise the point, hand-drawn graphs are used in this book. Some very elaborate pieces of complete rubbish are prepared using computers.

On the other hand, if you have access to a computer and graphics software, learning to use them is a good use of your time, particularly as it allows you to examine quickly a variety of relationships in your data and to determine which aspects merit graphical presentation. But don't get carried away with frills: once the graphs are plotted, for example, it may be faster to type or write by hand the axis labels or legends than to have the computer execute these steps for you.

When preparing graphs by hand, always use graph paper, although you can trace over the graph paper on white tracing paper if you wish. The most useful sort of graph paper has heavier lines at uniform intervals – at every ten divisions, for example, as shown in the graphs in Figures 20–24. These heavy lines facilitate the plotting of data and reduce eye-strain considerably, since every individual line need not be counted in locating data points.

By convention, the independent or explanatory variable is plotted on the *X*-axis (the horizontal axis, the abscissa) and the dependent or response variable is plotted along the *Y*-axis (the vertical axis, the

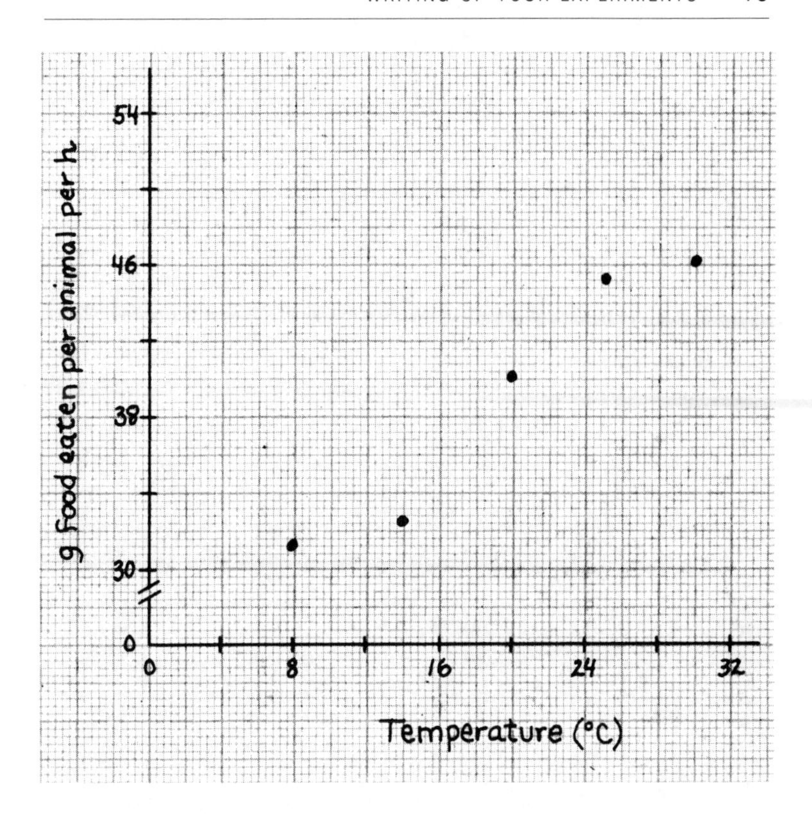

FIGURE 20.

Feeding rate of *Manduca sexta* caterpillars as a function of environmental temperature.

ordinate). For example, if you examined feeding rates as a function of temperature (Figure 20), you would plot temperature on the *X*-axis and feeding rate on the *Y*-axis; feeding rate *depends* on and responds to temperature, but temperature is not controlled by feeding rate and varies independently of feeding rate. Temperature is the independent variable and so is plotted on the *X*-axis.

It is good practice to label the axes of graphs beginning with zero. To avoid generating graphs with lots of empty, wasted space, breaks can be put in along one or both axes, as in Figures 17, 20 and 23–25). If a break had not been inserted in the *Y*-axis of Figure 20, for example, the

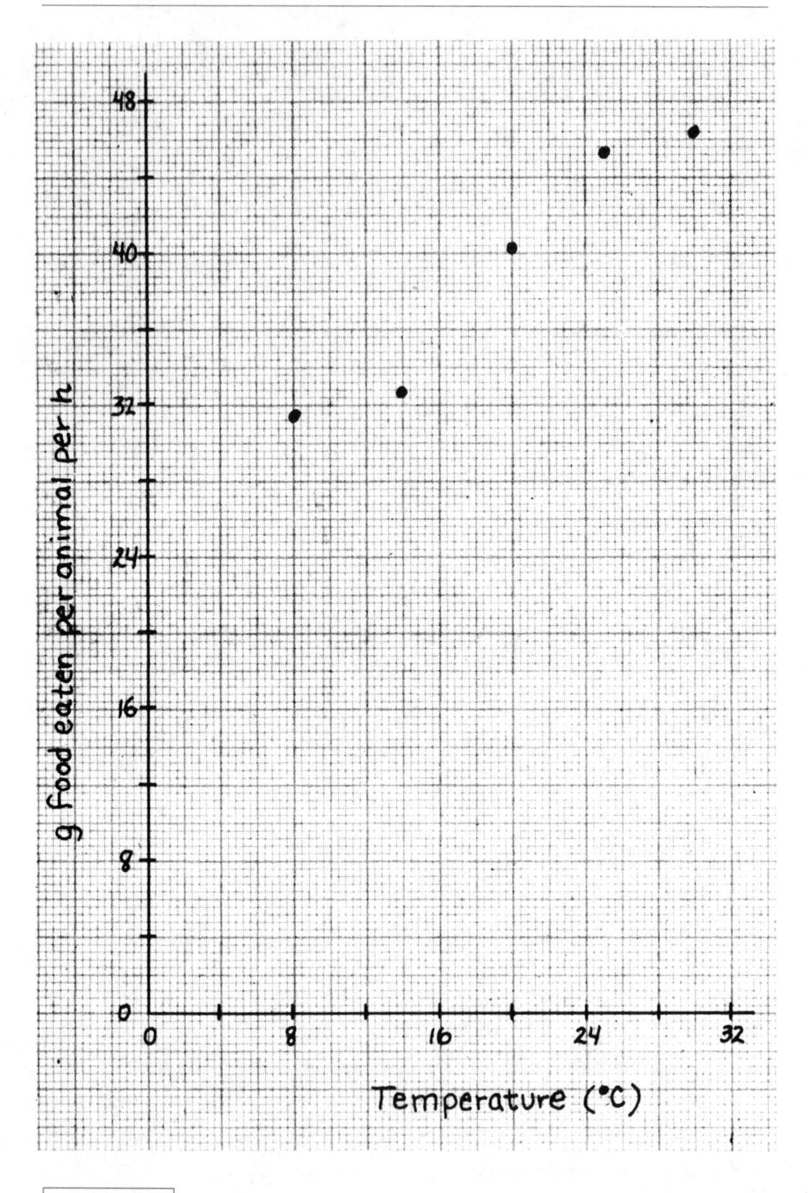

FIGURE 21.

Feeding rate of *Manduca sexta* caterpillars as a function of environmental temperature.

graph would have been less compact, as in Figure 21.

For some types of experiment, special kinds of graph paper – especially ones with non-linear axes – may be needed, such as log/log or log/linear paper, or probability paper or triangular co-ordinate paper. Staff may provide such paper, or tell you where you can get it. For example, if you measured the survival of yeast cells at different doses of UV light, you might get survivals of 100%, 9%, 1.1%. 0.08%, 0.009% and 0.0008%. These are difficult to display on a linear scale, but plotting the log of percentage survival against UV dose solves the problem of scale.

Connecting the Dots

After data points have been plotted, lines are often added to graphs to clarify trends in the data. It is especially important to add such lines if

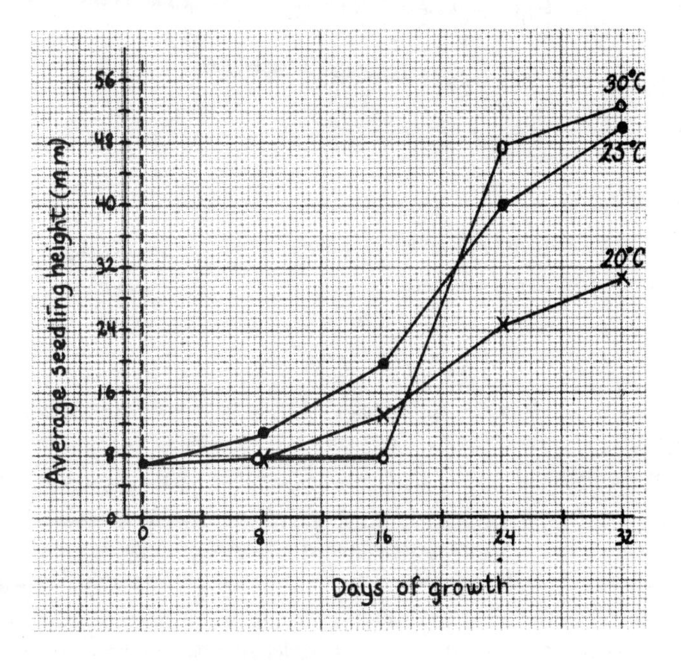

FIGURE 22.

Rate of seedling growth at three different temperatures.

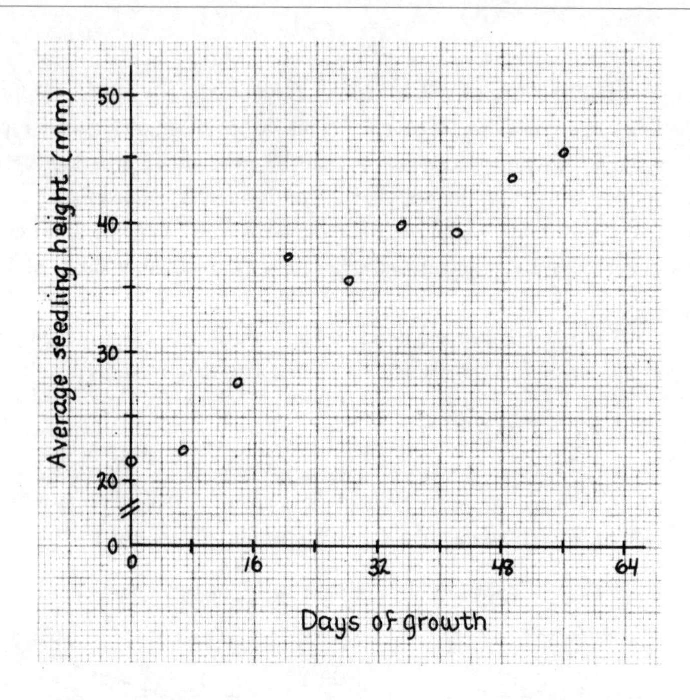

FIGURE 23.

Rate of seedling growth at 20°C. Fifteen to twenty seedlings were measured on each day of sampling.

data from several different treatments are plotted on a single graph, as in Figure 22 (previous page). This graph has been made easier to interpret by using different symbols for data obtained at each temperature. The zero point on the *X*-axis has been displaced to the right, to prevent the first data point from lying on the *Y*-axis where it might be overlooked (compare with Figure 23, where the first point does lie on the *Y*-axis).

In some cases it makes more sense to draw smooth curves than simply to connect the dots. For example, suppose we have monitored the increase in height of tomato seedlings over a period of time. Every week we randomly selected 15–20 seedlings to measure from the lab population of several hundred, so that different seedlings were usually

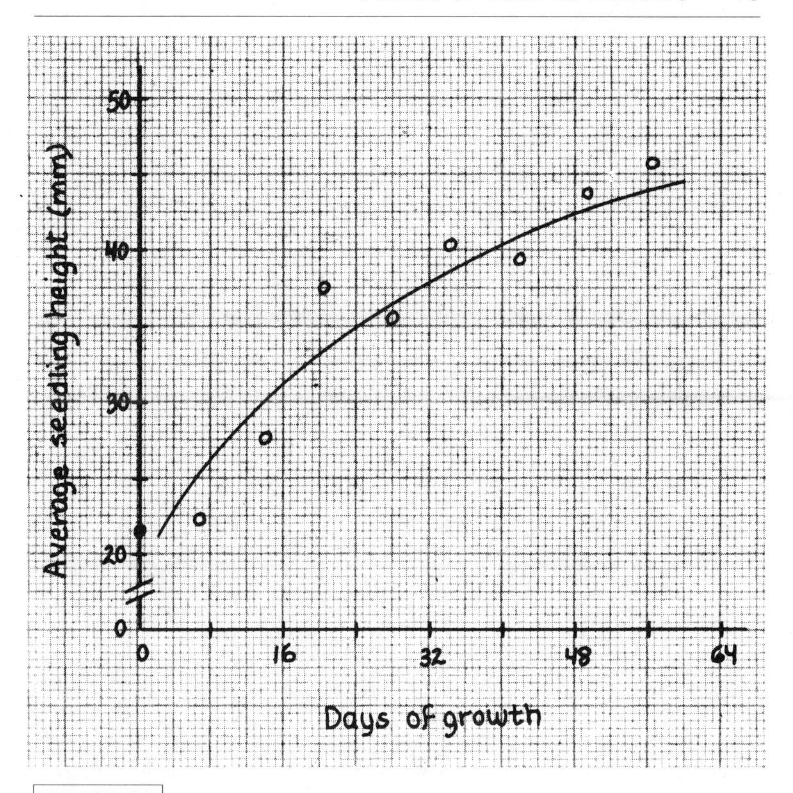

FIGURE 24.

Rate of seedling growth at 20°C. Fifteen to twenty seedlings were measured on each day of sampling.

measured at each sampling. After two months, the data were plotted as in Figure 23.

Connecting the dots would not be the most sensible way to reveal trends in the data of Figure 23, since we know that the seedlings did not really shrink between days 21 and 28 and between days 35 and 42. Simply connecting the points would suggest that shrinkage had occurred. This apparent decline in seedling height reflects the considerable variability in individual growth rates found within the population, as well as the fact that we did not measure every seedling in the population. The trend in growth is best revealed by drawing a smooth curve, as in Figure 24.

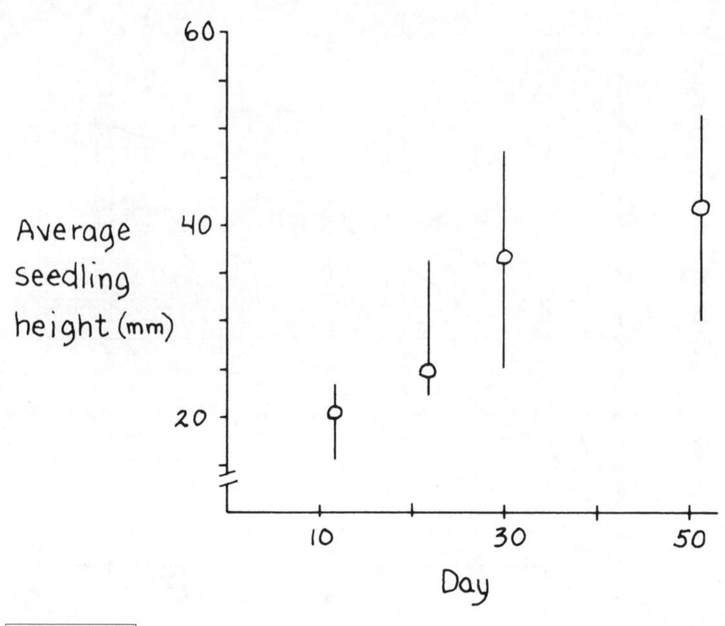

FIGURE 25.

Average height of tomato plant seedlings over a 50-day period. Vertical bars represent the range of heights.

When plotting average values, it is appropriate to include a graphic impression of the amount of variation present in the data by adding bars extending vertically from each point plotted (Figure 25). You may, for example, choose simply to illustrate the range of values obtained in a given sample. In Figure 25, for instance, the vertical bars extending from the point at day 30 indicate that although the average seedling height was about 37 millimetres (mm) on that day, at least one seedling in the sample was as small as 25 mm, and at least one seedling was as large as 46 mm. Alternatively, you may plot the standard deviation or standard error about the mean, in which case the vertical lines will extend equal distances above and below each point. Standard deviations and standard errors are reviewed in Appendix A, page 247. In your figure caption, be sure to indicate what you have plotted.

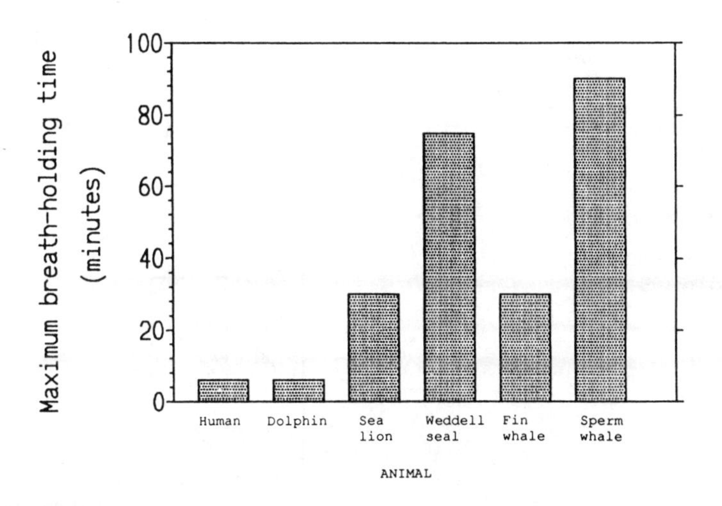

FIGURE 26.

Breath-holding abilities of humans and selected marine mammals. Data from Sumich 1992. *Biology of Marine Animals*, 5th ed. Wm. C. Brown, Publishers.

Bar Graphs and Histograms

When the independent variable along the *X*-axis is numerical and continuous, points can be plotted and trends can be indicated by lines or curves as we have seen in Figures 17–25. In Figure 20, for example, the *X*-axis shows temperature rising continuously from 0°C to 32°C, with each centimetre (cm) along the *X*-axis corresponding to a 4°C rise in temperature. Similarly, the *X*-axis of Figures 23–24 reflects the passing of time, from 0 to 60 days or so, with each cm along the *X*-axis reflecting eight additional days.

When the independent variable is non-numerical or discontinuous, or represents a range of measurements rather than a single measurement, the data are represented by bars, as shown in Figures 26–27. The *X*-axis of Figure 26 (a bar graph) is labelled with the names of different mammals. In contrast to the *X*-axes of Figures 17–25, the *X*-axis of Figure 26 does not represent a continuum; no particular quantity continually increases or decreases as one moves along the *X*-axis, and a line connecting

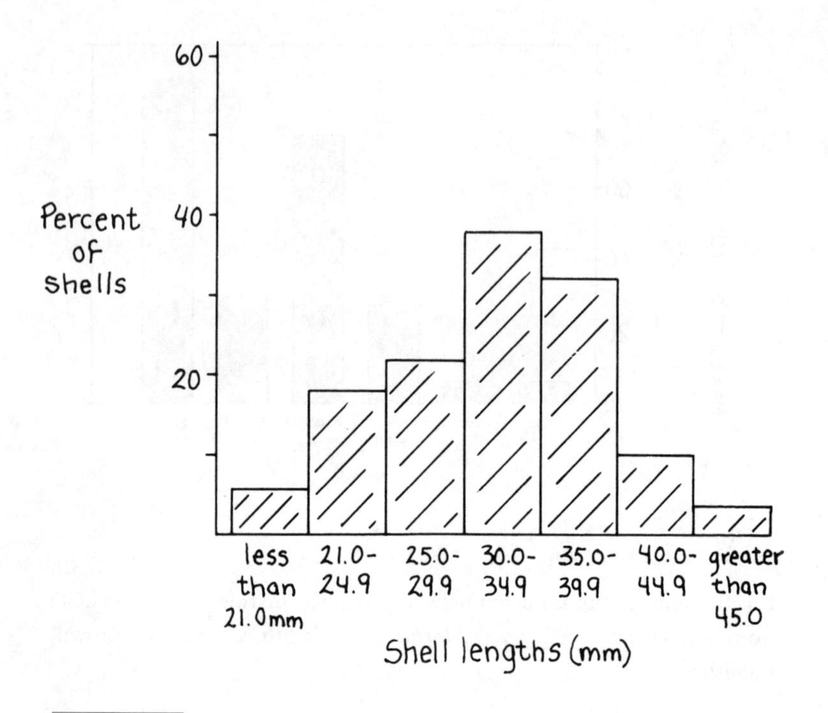

FIGURE 27.

Size distribution of snail shells *Littorina littorea* collected at Blissful Beach, Massachusetts, on August 15, 1989. Only living animals were measured. A total of 197 snails were included in the survey.

the data for sea lion and Weddell seal would be meaningless. In Figure 27 (a histogram), the data for shell length are numerical but are grouped together (for example, all shells 25.0–29.9 mm in length are treated as a single data point). Note that the magnitude of the size categories represented by the different bars varies; the left-most bar represents the percentage of shells found within a range of up to 21 mm in length, whereas each of the next several bars to the right represents the percentage of shells found within a range of only about 5 mm in length. The size range of shells represented by the bar at the extreme right side of the graph is unknown; all shells in this category exceeded 45 mm in length, but the graph does not indicate the size of the largest shell.

Preparing Tables

Tables should always be organised with data for a given characteristic being presented vertically rather than horizontally. Tables 3 and 4 present the same information but in different formats. Table 3 correctly places all information about a single species in one row, so that readers can view the information for each species by scanning from left to right, and can compare data among different species by scanning up and down a single column. Table 4 is incorrectly organised and more difficult to read. Like graphs, tables should be self-sufficient; note how much useful information the author has packed into the legend and column headings of Table 3.

YES:

Table 3. Characteristics of four snail populations sampled at Nahant, MA on October 13, 1985.

Species	Average shell length (cm)	Sample size	Average no. animals per m^2
Crepidula fornicata	1.63122 indiv.	32.1	
C. plana	1.01116	20.8	
Littorina littorea	0.87447	113.6	
L. saxatilus	0.4060	8.2	

NO:

Table 4. Characteristics of four snail populations sampled at Nahant, MA on October 13, 1985.

Species	Crepidula fornicata	C. plana	Littorina littorea	L. saxatilus
av. shell length (cm)	1.63	1.01	0.87	0.40
sample size	122 indiv.	116447	60	
aver. no. animals per m^2	0.40	20.8	113.6	8.2

If using word-processing, you should familiarise yourself with three kinds of Tab used for tables. A Left Tab lines up text with the left ends of words or numbers aligned, with the first character typed at the position of the Tab setting, and subsequent letters displayed to the right. If you were typing the key to Figure 19 (page 73), you would want the code letters A–F aligned under the letter 'o' of the heading 'Code', and the initial letters of the Genus names aligned under each other further to the right, so you would use two Left Tabs, perhaps six spaces apart, for these alignments. A Right Tab lines up text with the right-hand ends of words or numbers aligned; the first character typed is at the Tab position, with subsequent characters displayed to the left. In Table 3 (page 83), you could use a Right Tab to align each of the three sets of numbers at their right hand sides. If you used a Left Tab on the right-hand set of numbers, the number 113.6 would be displaced one place to the right and 8.2 would be displaced one place to the left as the initial numbers, not the final numbers, would line up. A Decimal Tab lines numbers up along their decimal points. Usually all numbers in a column are given to the same number of decimal places, so either a Right Tab or a Decimal Tab could be used, as for the left and right sets of numbers in Table 3, but if there were some special reason for different figures having different numbers of decimal places, then a Decimal Tab should be used to align them around the decimal point.

Laboratory and Field Drawings

In many areas of biology, biochemistry, medicine and related subjects, you will need to put laboratory or field drawings into your lab reports, project reports or theses. They can convey a very large amount of information very clearly. Drawings for lab reports are usually done on plain, unlined white paper, in pencil, while those for theses and journal articles are usually in Indian ink or other drawing inks. For fine detail, a moderately hard pencil such as 2H is useful. For lab reports, use original drawings done in the lab, without redrawing them, when inaccuracies could occur. Copying drawings from textbooks is plagiarism; if your drawing is not original, state its source.

Each drawing must be large enough to show detail, typically half a page or more. It must have a full explanatory caption, including the organism, the part shown, the viewpoint from which the drawing was made, and perhaps what features are illustrated. For example: '*Drosophila melanogaster*, adult female of wild-type stock C, whole fly, dorsal view, to show features differing between the sexes'. If the whole report is on the same organism, there is no need to repeat the name of the species in each caption. Another drawing in the same report might be captioned: 'Adult male, mutant stock A, with white eyes, bar eyes and miniature wings, side view of head to show eye shape'.

All drawings must have a magnification or scale bar, *e.g.* life size; ×400; the fly is 3 mm long; scale bar represents 10 μm. Use a ruler to measure the organism and your drawing to calculate the magnification of macroscopic objects. For example, if your drawing is 15 cm long for a fly 3 mm long, the actual magnification is ×50, even if the drawing was done using ×10 magnification on the dissecting microscope. In this case, it is more accurate to measure your fly by aligning your ruler with the fly under the microscope, than to measure such a small length using the naked eye. For microscopic objects, calibrate your microscope separately for each magnification, using a micrometer slide which has a calibrated line with divisions an accurate distance apart, as stated on the slide, *e.g.* 'large divisions 0.10 mm, small divisions 0.01 mm'. For rough purposes, calibrate the field width at each magnification; if under a ×10 objective and ×10 eyepieces you can see 13 large divisions on such a micrometer slide, then your field width is 1.3 mm at ×100 magnification. If you could see 13 small divisions with the ×100 objective, then your field width would be 0.13 mm at ×1000, so that a specimen which was one-third as long as your field width at ×1000 would be 0.043 mm long, or 43 μm.

For more accurate measurements under the microscope, use an eye-piece graticule to measure the specimen in eye-piece units, then calibrate the eye-piece units in absolute lengths using the micrometer slide. For really accurate, fast measurements, use a filar (moving hair) micrometer eye-piece if one is available. For electron micrographs, ask the staff how to get absolute lengths or magnifications.

Don't forget that magnifications can change from negative to print, depending on the degree of enlargement used. Special instruments for drawing from the microscope, such as a camera lucida, may be available, and you may be able to take photomicrographs if the equipment is provided.

It is good practice to state the microscope magnification used for making a drawing, but be sure to give the total magnification, including the eye-pieces, not just the objective lens magnification. Do remember that you must give the specimen-to-drawing magnification as well, not just the microscope magnification, as you could make a large or small drawing from the same microscope view. If you used any special illumination, lenses or condenser, these should be mentioned, *e.g.* dark-field illumination, phase contrast, oil-immersion lens.

Your drawings should be carefully labelled to show the features of interest, with a line drawn along a ruler from the part of the drawing to your label. Labels are usually in lower case letters, not capitals, and can be in pencil or ink for most lab reports, unless specified by your lecturer. For theses and journal articles, machine-printed labelling is usually required: for journals, check the instructions to authors for that journal. Students often do not use enough labels for the features of interest, and often use too few drawings, or drawings of too small a size.

Bear in mind the purpose of each drawing as you make it. For example, if you are illustrating the differences between the sexes in fruit flies, your drawing should concentrate on those features which differ between the sexes, such as overall size, abdomen pigmentation and shape, genitals, and the sex combs on the front legs of the male. There would be no need to draw all the abdominal bristles or give details of wing venation or mouth parts, as these do not differ between the sexes. Most biological drawings are of outlines only, without shading, but to show, for example, the difference between black bands and pale ones on a snail, you would shade in the black bands and label them appropriately. Colouring is not usually used in lab report drawings, as natural colours are hard to reproduce accurately. Labelling with words is usually sufficient, as in: 'bright gold bacterial colony, *Staphylococcus aureus*'.

If you are making a schematic diagram, rather than a drawing, say of the life history of a plant, you may have to draw different parts on a quite different scale, *e.g.* the mature plant and the cells inside the ovule. Be sure then to indicate that different parts are not drawn on the same scale.

Giving the Results in Words

One-sentence Results sections are common in student reports: 'The results are shown in the following tables and graphs.' However, *common* does not mean *acceptable.* You must use words to draw the reader's attention to the key patterns in your data. But do not simply redraw the graphs in words, as in this description of Figure 22 (page 77):

> At 20°C, the seedlings showed negligible growth for the first 8 days of study. However, between days 8 and 16, the average seedling grew nearly 5 mm, from about 8 mm to about 13 mm. Growth continued over the next 16 days, with the seedlings reaching an average height of 24 mm by day 24, and 30 mm by day 32.

Let the graph do this work for you; your task is to summarise the most important trends displayed by the graph. For example, you might write:

> Temperature had a pronounced effect on seedling growth rates (Figure 22). In particular, seedlings at 25°C consistently grew more rapidly than those at 20°C. . .

Let us apply these principles to the caterpillar study (page 63). First, is there anything about the general response of the animals worth drawing attention to? You might write:

> All of the caterpillars were observed to eat throughout the experiment.

More likely, living things behaving as they do, you will put something like:

> One of the animals offered diet A and two of the animals offered diet B were not observed to eat during the three-hour experiment; the results from these animals were therefore excluded from analysis.

Such a decision to exclude data from further analysis is fine as long as you give a good objective reason for it. You cannot exclude data simply because it violates a trend that would otherwise be apparent, or because the data contradict a favoured hypothesis. Next, go back to your initial list and reword each question as a statement. For example, the first question posed earlier, on page 67 ('Do the caterpillars feed at different rates on the different diets?'), might be reworded as:

> Caterpillars generally fed at faster rates on diet A than on diet B.

In scientific writing, statements of fact should be backed by evidence. Here you can support the statement with a reference to the appropriate figure or table.

> Caterpillars generally fed at faster rates on diet A than on diet B (Table 2).

Readers can then look at Table 2 and decide whether they see the same trend. Sometimes you may add a specific example from your data to support your statement, but that is unnecessary here. One sentence and a table say it all. If you follow this procedure for each question on your list, your Results section will be complete. The written part will generally be fairly short.

Note that the statement about the rates at which caterpillars fed does not mention the term *significant*; it does not say, 'Caterpillars fed at significantly faster rates on diet *A* than on diet *B*.' Using the term *significant* implies that you have subjected the data to an appropriate statistical test to determine that the differences observed are substantial enough to be convincing, not just due to chance variation. Do not write about significant differences, or the lack of them, unless you have conducted such a test. An introduction to the use of statistics in interpreting results begins on page 114.

Note the use of the past tense in the statement about caterpillar feeding rates:

> Caterpillars generally fed at faster rates on diet A.

This statement is quite different from the following one, in which the present tense is used:

> Caterpillars feed at faster rates on diet A.

By using the present tense, you would be making a broad generalisation extending to all caterpillars, or at least to all caterpillars of the species tested. Before one can make such a broad statement, the experiment must be repeated many times, and similar results must be obtained each time; after all, the writer is making a statement about all caterpillars under all conditions. By sticking with the past tense, however, you are clearly referring only to the results of your study.

Writing About Numbers

According to the Council of Biology Editors, you should use numerals when presenting percentages, decimals, magnifications, and abbreviated units of measurement: 25%, 1.5 times greater, ×15 magnification, 0.7 g, 18 ml. In most other situations, use words for numbers zero to nine, inclusive, and numerals for larger whole numbers: seven seedlings, 15 nucleotides, 247 protozoans. When using a series of numbers at least one of which is greater than nine, use numerals for all; for example:

> We added 6 drops to the first flask, 9 drops to the second and 12 drops to the third.

Writing About Negative Results

An experiment that was correctly performed always 'works'. The results may not be what you had expected, but this does not mean that the experiment has been a waste of time. If biologists threw away their data every time something unexpected happened, we would rarely learn anything new. The data you collect are real; it is the interpretation that is open to question. Always treat your data with respect. The lack of a trend or the presence of a trend contrary to expectation is itself a story worth telling.

Think Quantitatively

Where relevant, express your findings quantitatively rather than qualitatively. It looks and sounds more scientific, and helps you and the reader to grasp the magnitude and the importance of any differences found. Don't write, 'Exon 2 of the acetylase enzyme is somewhat larger than exon 3', if your messenger RNA sequencing data show that: 'Exon 2 of the acetylase enzyme is 2.5 times longer than exon 3.'

IN ANTICIPATION

Much of the work in putting together a good lab report goes into preparing the Results section. You can save yourself considerable effort and frustration by planning ahead before you enter the lab to do the experiment. Be prepared to record your data in a format that will enable you to make your calculations easily. For the caterpillar experiment referred to previously, you would be well ahead by coming to the lab with a data sheet set up like the one in Figure 28 (page 91). With this sheet, the data are recorded in the *X* areas; the blank spaces will be filled in later, as you make your calculations. If possible, leave a few blank columns at the right, to accommodate unanticipated needs discovered as you record or analyse your data. In introductory lab exercises, students are often provided with data sheets already set up in a useful format. It is worth taking a careful look at these data sheets in order to understand how they are organised and why they are as they appear; in more advanced lab courses, you will be responsible for organising your own data sheets. As mentioned earlier, always follow any number you write down with the appropriate units, such as mg (milligrams), cm (centimetres), or mm/min (millimetres per minute, often written as $mm.min^{-1}$).

CITING SOURCES

The next sections to prepare are the Discussion and the Introduction, in that order. In both sections you make statements of fact that require

Date and time started: ____________

Date and time ended: ____________

Caterpillar No.	Diet	Caterpillar wt. (g) Initial	Caterpillar wt. (g) Final	Weight change (g)	Food wt. (g) Initial	Food wt. (g) Final	Food wt. change (g)	Feeding rate g eaten/caterp./h
x	x	x	x		x	x		

FIGURE 28.

Sample format for a laboratory data sheet.

support, often from written sources. As stated in Chapter 1, most statements of fact or opinion must be supported with a reference to their sources; this applies particularly to scientific papers and theses, but is not always expected in simple lab reports. Consult the lecturer concerned if in doubt. Here are a few general rules to follow when backing up factual statements in your report.

1. **Cite the reference in the text.** Don't use footnotes. In most papers, references are cited directly in the text, by author and year of publication, as in the following example.

   A variety of organic molecules are commonly used to maintain or adjust the osmotic concentration of intracellular fluids (Hochachka and Somero, 1984; Schmidt-Nielsen, 1990).

   Some journals connect the names of two authors with 'and', as in that example, but others use an ampersand (&). A few journals use numbered superscripts (-[1]) in the text, then list the references at the end of the paper in numerical order of the superscripts, instead of in alphabetical order of authors' surnames. Here are two parts of sentences from Lamb & Wickramaratne, *Nature*, 1975, **258**, 645–646, where the first part mentions the author's name but the second part does not mention names at all, just superscript numbers. Readers then have to turn to [4] and [5] in the list of references if they wish to know whose work is referred to, which is not good practice, but saves space:

   Leblon[1] showed a clear relation between. . . . Because mutations may differ in their conversion responses to genetic and environmental factors[4,5], different mutants. . . .

   When more than two authors have collaborated on a single publication, a shortcut is standard practice:

   A mutation is defined as any change occurring in the nitrogenous base sequence of DNA (Tortora *et al.*, 1982).

The *et al.* is an abbreviation for *et alii*, meaning 'and others'. The words are underlined or italicised even when abbreviated, because they are in a foreign language, Latin; underlining tells a printer to set the designated words or letters in italics. Note that the full stop (period) follows the closing parenthesis, since the reference is part of the sentence. The authors' names can also be incorporated directly into a sentence:

> Holliday (1990) has reviewed the literature on the control of gene expression during embryonic development.

You can cite your lab manual by its author (*e.g.* 'B. Stewart, 1993') or as follows:

> Preparation of buffers and other solutions is described elsewhere (Biology 1 Laboratory Manual, University College London, 1993).

If your information comes from a lecture or from a conversation with a particular individual, support your statement as follows:

> California grey whales migrate up to 18,000 km yearly (Professor J. Tashiro, personal communication).

2. **Be concise in citing references.** Avoid writing,

> In his classic work, *The Biology of Marine Animals*, published in 1967, Colin Nicol reviewed the literature on invertebrate bioluminescence.

Instead, write,

> Invertebrate bioluminescence has been reviewed by Nicol (1967).

3. **Ideally, cite only those sources you have actually read.** Don't list references simply to add bulk; your lecturer is justified in expecting you

to be able to discuss any material you cite. However, you may occasionally have to cite a reference that you have not actually read, perhaps because it is not available. For example, when the results reported by Smith (1964) are cited in a book written by Jones (1983), and you have read only the work by Jones. Your citation should then read, '(Smith, 1964; as cited by Jones, 1983)'. Let Jones take the blame if he or she has misinterpreted something.

THE DISCUSSION SECTION

In this section, you must interpret your results in the context of the specific questions you set out to address and in the context of any relevant broader issues that have been raised in lectures, textbook readings, previous coursework and perhaps library research. Consider the following issues:

1. What did you expect to find, and why?
2. How did your results compare with those expected?
3. How might you explain any unexpected results?
4. How might you test these potential explanations?

If your results coincide exactly with those expected from prior knowledge, your Discussion section will be rather short. Such a high level of agreement is rarely obtained. Indeed, high degrees of variability characterise many aspects of biology. An experiment may need to be repeated many times, with large sample sizes, before clear trends emerge. This point is discussed further in the section on statistical analysis beginning on page 114. A short paper in a biological journal may well represent years of work by several competent, hard-working individuals. Even the simplest questions are often not easily answered. Nevertheless, every experiment that was carried out properly tells you *something*, even if it is not what you intended to find out.

Expectations

State your expectations explicitly and back your statements up with a reference if possible. Scientific hypotheses are not simply random guesses. Your expectations must be based on facts, not opinions; these facts could come from lectures, lab manuals or handouts, textbooks, or other sources. In discussing a study on the effectiveness of different wavelengths of light in promoting photosynthesis, for example, you might write something like:

> All wavelengths of light are not equally effective in promoting photosynthesis: green light is said to be especially ineffective (Smith and Jones, 1981). This is because green light tends to be reflected rather than absorbed by plant pigments, which is why most plants look green (Smith and Jones, 1981). Our results supported this expectation. In particular. . . .

Alternatively, a discussion section might profitably begin:

> The results of our experiment failed to support the hypothesis (Smith and Brown, 1989) that caterpillars of Manduca sexta reared on one uniquely flavoured diet will prefer that diet when subsequently given a choice of foods.

Here we have managed to state our expectations and compare them with our results in a single sentence. In both cases, we have begun our discussion on firm ground – with facts rather than opinions.

Explaining Unexpected Results

When results refuse to meet expectations, students commonly blame the equipment, the staff, their lab partners or themselves. Generally, more scientifically interesting possibilities than experimenter incompetence are the culprits. Don't be too hard on yourself. Take another look at the list of factors you wrote down when beginning to work on your Materials and Methods section. Could any of these factors be sufficiently different from the normal or standard conditions under which the experiment is performed to account for the differences? Were any of the conditions under which your experiment was performed substantially different from those assumed in the instruction manual? If you discover no obvious differences in the experimental conditions, or if the differences cannot account for your results, include this point in your report.

> The discrepancy in results cannot be explained by the unusually low temperature in the laboratory on the day of the experiment, since the control animals were subjected to the same conditions and yet behaved as expected.

If potentially important differences are noted, put this ammunition to good use.

> In prior years, these experiments have been performed using species X. It is possible that species Y simply behaves differently under the same experimental conditions.

Note that the writer does not *conclude* that species *X* and species *Y* behave differently; the writer merely *suggests* this explanation as a possibility. Always be careful to distinguish possibility from fact. Suggesting a logical possibility won't get you into any trouble. Stating your idea as though it were a proven fact, on the other hand, is a bad scientific error. Continue your discussion by indicating possible ways in which the differences in behavioural responses might be tested. For example:

> This possibility can be examined by simultaneously exposing individuals of both species to the experimental conditions. If species X behaves as expected, and species Y behaves as it did in the present experiment, then the hypothesis of species-specific behavioural differences will be supported. If species X and species Y both respond as species Y did in the present study, then some other explanation will be called for.

Continue in this vein, evaluating all the reasonable, testable possibilities you can think of. A lecturer enjoys reading these sorts of analyses because they indicate that students are thinking about what they've done.

Notice that in the preceding paragraph the writer did not say, 'If species *X* behaves as expected, and species *Y* behaves as it did in the present experiment, then the hypothesis will be supported.' This writer remembers rule number 7 (page 6): Never make the reader turn back. Notice, too, that the writer does not put, '. . . then the hypothesis will be true' or '. . . then the hypothesis will be proven.' Experiments cannot *prove* anything; they can only support or disagree with hypotheses. Our interpretations of phenomena may make excellent sense based upon what we know at the moment, but those interpretations are not necessarily correct. New information often changes our interpretations of previously acquired data.

Analysis of Specific Examples

Example 1

In this study, tobacco hornworm caterpillars were raised for four days on one diet, and then tested over a three-hour period to see if they preferred that food when given a choice of diets.

Student Report

The data indicate that the choice of food was not related to the food upon which they had been reared. These data run counter to the hypothesis (Back and Reese, 1976) that hornworms are conditioned to respond to certain specific foods. Only one set of data out of the four gave any indication of a preference for the original diet, and that indication was rather weak.

There are many possible explanations for data so contrary to previous experimental results. Inexperience of the experimenters, combined with the fact that three different people were recording data about the worms, may account for part of the error. Keeping track of many worms and attempting to interpret their actions as having chosen a food or having merely been passing by may have proven to be too much for first-time worm watchers. The mere fact that each of three people will interpret actions differently and will have somewhat different methods of recording information introduces bias into the data.

Analysis

This Discussion section starts out well, with a comparison between results expected and results obtained. The hypothesis being discussed is clearly stated, and a supporting reference is given. Moreover, the student recognises that 'data are' rather than 'data is'. (However, the student persists in calling the animals worms rather than caterpillars; the term *worms* usually refers to annelids, whereas these animals are arthropods.) The next paragraph, however, betrays a total lack of confidence in the data obtained; the results could not possibly have turned out this way unless the researchers were incompetent, writes the student. Although inexperience can certainly contribute to suspicious results, are there no other possible explanations? Does it really take years of training to determine whether a caterpillar is eating food *A* or food *B*? Compare this report with the one in the next example. This Discussion deals with the same experiment. In fact, the two students were laboratory partners.

Example 2

Student Report

Contrary to expectation, the results suggest that caterpillars show no preference in the diet they touched first and the diet they spent the most time feeding on. This unexpected finding may be due to the fact that the caterpillars were not reared on the original diets for a long enough period of time to acquire a lasting preference. They ate only the diets they were reared on for four days, whereas the laboratory handout had suggested a pre-feeding period of 5–10 days (Chew and Pechenik, 1988). This possibility may be tested by performing the same experiment but varying the amount of time that the caterpillars are reared on the original diets. Such an experiment would determine whether there is a critical time that caterpillars should be reared on a particular diet before they will show a preference for that diet. Another possible explanation for the results obtained may be that the caterpillars used were very young, weighing only 3–6 mg. Finally, this experiment lasted only 3 hours. Perhaps different results would have been obtained had the organisms been given more time to adjust to the test conditions. It would be interesting to run the identical experiment for a longer period of time, such as 10–12 hours.

The author of this report produced a paper that clearly indicates thought. Which report do you suppose received the higher mark?

Example 3

In this experiment, several hundred millilitres (ml) of filtered pond water was inoculated with a small population of the ciliated protozoan, *Paramecium multimicronucleatum,* and then distributed among three small flasks. Over the next five days, changes in the numbers of individuals per ml of water in each flask were monitored.

Student Report

The large variation observed between the groups of three replicate populations suggests that the experimental technique was imperfect. The sampling error was high because it was difficult to be precise in counting the numbers of individuals. Some animals may have been missed while others were counted repeatedly. More accurate data may be obtained if the number of samples taken is increased, especially at the higher population densities. In addition, more than three replicate populations of each treatment could be established. Finally, extremely precise microscopes and pipettes could be used by experienced operators in order to reduce sampling error.

Analysis

This writer, like the writer of Example 1, starts out by assuming that the experiment was a failure and spends the rest of the report making excuses for this failure. The quality of the microscopes was certainly adequate to recognise moving objects, and *P. multimicronucleatum* was the only moving organism in the water. The author is grasping at straws. If the author had more confidence in his or her abilities, the report might have been very different. Isn't there some chance that the experiment was performed correctly? Lacking confidence in the data, the student looked no further even though he or she actually had access to data that would have allowed several of the stated hypotheses to be assessed. On each day several sets of samples were taken from each flask and each set gave similar estimates for numbers of organisms per ml; this consistency of results suggests that the variation in population density from flask to flask was not due to experimenter incompetence. In addition, although the student stated correctly that larger sample sizes would have been helpful, he or she should have supported that statement with additional analysis of the data. Fifteen drops were sampled from each flask for each set of samples. The student could have calculated the mean number of individuals in the first 3 drops, the first 6 drops, the first 9 drops, the first 12 drops, and then the full 15 drops, to see how estimates of population size changed as the sample size was increased. If the student had done this, he or she would probably have found that larger sample sizes are especially important when population size is least dense. (Why might this be so?)

Example 4

In this last study, a group of students went seining for fish in a local pond. Every fish was then identified by species, so that the number of fish of each species could be determined. It turned out that 91 per cent of the fish in the sample of 73 individuals belonged to a single species. The remaining fish were distributed among only two additional species.

Student Report

I find the small number of species represented in our

> sample surprising, since the pond is fed by several streams that might be expected to introduce a variety of different species into it, assuming that the streams are not polluted. The lab manual states that 12 fish species have been found in the adjacent streams. It appears that the conditions in the pond at the time of our sampling were especially suitable for one species in particular of all those that most likely have access to it. Perhaps the physical nature of the pond is such that the number of niches is small, in which case competition would become very keen; only one species can occupy a given niche at any one time (Smith, 1958). The reproductive pattern of the fishes might also contribute to the observed results. Possibly Lepomis macrochirus, the dominant species, lays more eggs than the others, or perhaps the juveniles show better survival.
>
> Another possible explanation for our findings might lie in the fact that we sampled only the perimeter of the pond, since our seining was limited to a depth of water not exceeding the heights of the seiners. The species distribution could be very different in the middle of the lake at a greater depth.

Analysis

This excerpt from the Discussion section demonstrates that a little thinking goes a long way. The student did not require much specialised knowledge to write this Discussion, only a bit of confidence in the data. Another student might well have written,

> Most likely, the fish were incorrectly identified; more species were probably present than could be recognised by our inexperienced team. It is also possible that the net had a large tear, which let most of the species escape. I didn't notice this rip in the fabric, but my glasses were probably dirty, and then again, I'm not very observant.

THE INTRODUCTION

The Introduction establishes the framework for the entire report. In this section you briefly present background information that leads to a clear statement of the specific issue or issues that will be addressed in the remainder of the report. By the time you have completed writing the Materials and Methods, Results, and Discussion sections of your lab report, you should know what these issues are. In one or two paragraphs you must explain why the study was undertaken. Every topic that follows this section should be anticipated clearly in the Introduction. Here is a simple and satisfactory Introduction, written by a first-year undergraduate for her first university practical, which was a multi-purpose set of experiments.

> This set of simple classroom experiments were [was] designed to show how micro-organisms in various places, such as the air and on the skin, may contaminate experiments, and how sterile techniques, e.g. flaming, can be used to overcome contamination of pure cultures by such microbes. These techniques are of great importance in the laboratory to prevent spread of potentially pathogenic organisms, and to enable growth of pure cultures when needed.

Stating the Question

Even though the statement of questions posed, or issues addressed, generally concludes the Introduction of a report, it is helpful in writing this section of the paper to deal with this issue first. What *was* the point of this study? Write the following words: 'In this study' or 'In this experiment'. Then complete the sentence as specifically as possible. Three examples follow.

> In this study, the oxygen consumption of mice and rats was measured in order to investigate the relationships between metabolic rate, body weight, and body surface area.

> In this study, we collected fish from two local ponds and classified each fish into its proper taxonomic category.

> In this experiment, we asked the following question: do the larvae of Manduca sexta prefer the diet upon which they have been reared when offered a choice of diets?

Note that each statement is in the past tense since the students are describing completed studies. The strong points of these statements are best revealed by examining a few unsatisfactory ways to complete sentences dealing with the same material:

> In this study, we measured the metabolic rate of rats and mice.

> In this study, we made a variety of measurements on fish.

> In this experiment, the feeding habits of Manduca sexta larvae were studied.

Each of these unsatisfactory statements is vague, suggesting that their writers are in the dark about what they have done. Be specific. In one sentence, you must come fully to grips with your experiment or study. There *was* some point to the time that you were asked to spend in the lab; find it. Many experiments are multi-purpose, to introduce you to new techniques, as well as to study biological problems. So if an experiment introduces you to, say, gel electrophoresis of DNA fragments, then mention that in the Introduction.

An Aside: Studies Versus Experiments

An experiment involves manipulating something, such as an organism, an enzyme, or the environment, in a way that will permit specific hypotheses to be tested. Containers of protozoans in pond water could be distributed among three different temperatures, for example, to test the influence of temperatures on reproductive rates. An experiment may be conducted in the laboratory or in the field.

It is permissible to refer to experiments as 'studies', but not all studies are 'experiments'. Some exercises only require you to collect, observe, enumerate or describe. You should avoid referring to such studies as 'experiments'; where there are no manipulations, there are no experiments. You might, for example, collect and identify insects from light fixtures in different locations within the Biology building, enabling you to examine the distribution of insects within the building. Or you might document the depth to which light penetrates into various areas of a lake and then correlate that information with data on the distribution of aquatic plants in the different areas. In each case, you should refer to a study, not an experiment.

Providing the Background

Having posed in a single sentence the question or issue that was addressed, it is relatively easy to fill in the background needed to understand why the question was asked. A few general rules should be kept in mind:

1. **Ideally, back all statements of fact with a reference to your textbooks, lab manual, outside reading, or lecture notes.** Within the text, give the author of the source and the year of publication as previously described (page 92).

2. **Define specialised terminology.** By defining specialised terms in your own words, you can help yourself, as well as convincing the lecturer that you know what these words mean. As always, write with your future self in mind as the audience; write an introduction you will be able to understand five years from now. The following three sentences obey this and the preceding rule.

    A number of caterpillar species are known to exhibit induction of preference, a phenomenon in which an organism develops a preference for the particular flavour on which it has been reared (Jones and Smith, 1983).

Molluscs are common inhabitants of the intertidal zone, that region of the ocean lying between the high and low tide marks (Jones and Peters, 1983).

The development of mature female gametes, a process termed oogenesis, is regulated by changing hormonal levels in the blood (Browder, 1984; Wilson, 1979).

3. **Never set out to prove, verify, or demonstrate the truth of something.** Rather, set out to test, document or describe. It is important to keep an open mind when you begin a study and when you write up the results. It is not uncommon to repeat someone else's experiment and get a different result. Responses will differ with species, time of year, and other, often subtle, changes in the conditions under which the study is conducted. To show that you had an open mind when you undertook your study, you would want to revise the following sentences.

In this experiment we attempted to demonstrate induction of preference in larvae of Manduca sexta.

This study was undertaken to verify the description of feeding behaviour given for Manduca sexta by Jones (1903).

This experiment was designed to show that pepsin, an enzyme promoting protein degradation in the vertebrate stomach, functions best at a pH of 2 (Jones and Smith, 1983).

The first example might be modified to read,

> In this experiment, we tested the hypothesis that young caterpillars of Manduca sexta demonstrate the phenomenon of induction of preference.

How would you modify the other two examples to show that you approached the studies without prejudice?

4. **Be brief.** Include only the information that directly prepares the reader for the statement of intent, which will appear at the end of the Introduction as discussed above. If, for example, your study was undertaken to determine which wavelengths of light are most effective in photosynthesis, there is no need to describe the detailed biochemical reactions of photosynthesis. As another example, consider these sentences from a report on a study of induction of preference. Caterpillars were reared on one diet for five days and tested later to see if they chose that food over foods they had never before experienced.

> In this experiment, we explored the possibility that larvae of Manduca sexta could be induced to prefer a particular diet when later offered a choice of diets. The results of this experiment are significant because induction of preference is apparently linked to (1) the release of electrophysiological signals by sensory cells in the animal's mouth and (2) the release of particular enzymes, produced during the period of induction, that facilitate the digestion and metabolism of secondary plant compounds (lab handout, 1992).

The entire last sentence does not belong in the Introduction. The work referred to above was a simple behavioural study; students did not make electrophysiological recordings, nor did they isolate and characterise any enzymes. Although a consideration of these two topics might profitably be incorporated into a discussion of the results, these issues should be excluded from the Introduction because they do not explain why the study was undertaken. Include in your Introduction only information that directly prepares the reader for the final statement of intent. You might, on a separate piece of paper, jot down other ideas that occur to you for possible use in revising your Discussion, but if they don't make a contribution here, don't let them intrude on your Introduction.

5. **Write an introduction for the study that you ended up doing.** Sometimes it is necessary to modify a study for a particular set of conditions, so that the observations actually made no longer relate to the questions originally posed in your lab schedule. For example, the pH meter might not have been working, so perhaps the experiment you actually performed dealt with the influence of temperature, rather than pH, on enzymatic reaction rates. In such an instance, you would make no mention of pH in your Introduction section, since the work you ended up doing dealt only with the effects of temperature.

   The following paragraph satisfies all the requirements of a valid introduction. It is brief but complete . . . and effective.

> It is well known that plants are capable of using sunlight as an energy source for carbon fixation (Jones and Smith, 1983). However, all wavelengths of light need not be equally effective in promoting such photosynthesis. Indeed, the green colouration of most leaves suggests that wavelengths of approximately 550 nm are reflected rather than absorbed, so that this wavelength would not be expected to produce much carbon fixation by green plants. During photosynthesis, oxygen is liberated in proportion to the rate at which carbon dioxide is fixed (Jones and Smith, 1983). Thus relative rates of photosynthesis can be determined either by monitoring rates of oxygen production or by monitoring rates of carbon dioxide uptake. In this experiment, we monitored rates of oxygen production under different light conditions, to test the hypothesis that wavelengths differ in their ability to promote carbon fixation by Elodea canadensis.

DECIDING ON A TITLE

In many cases for lab reports, the title will be given in your schedule. If it is not given, use a title that demonstrates to your lecturer that you have understood the point of the exercise. A good title summarises as specifically as possible what lies within the Introduction and Results sections of the report. When you are a professional biologist, in the real world of publications, your article will vie for attention with articles written by many other people; the busy potential reader of your paper will often glance at the title of your report and promptly decide whether to stay or move on. The more revealing your title is, the more easily your potential audience can assess the relevance of your paper to their interests. A paper that delivers something other than what is promised by the title can lose you good will when read by the wrong audience and may be overlooked by the audience for which the paper is intended. Indeed, many potential readers will miss your paper entirely, since indexing services such as *Biological Abstracts* and *Current Contents* use key words from a paper's title in preparing their subject indexes.

Here is a list of mediocre titles, each followed by one or two more revealing counterparts:

No: Metabolic rate determinations
Yes: Exploring the relationship between body size and oxygen consumption in mice

No: Plankton sampling in a small pond
Yes: Species composition of the spring zooplankton of Chew Lake, Avon

No: 1) Measuring the feeding behaviour of caterpillars
2) Eating habits of *Manduca sexta*
3) Food preferences of *Manduca sexta* larvae
Yes: 1) Measurements of feeding preferences in tobacco hornworm larvae (*Manduca sexta*) reared on three different diets

2) Can larvae of Manduca sexta (Arthropoda: Insecta) be induced to prefer a particular diet?

No: Effects of pollutants on sea urchin development
Yes: Influence of Cu^{++} on fertilisation success and gastrulation in the sea urchin Strongylocentrotus purpuratus

The original titles are too vague to be compelling. In scientific papers, avoid unnecessary words in the title, such as 'Studies on. . .' or 'Experiments on. . .'.

WRITING AN ABSTRACT

The abstract, which is usually only required in theses or scientific papers, is placed at the beginning, immediately following the title page. Yet it should be the last thing you write, since it must completely summarise the essence of your report: why the experiment was undertaken; what problem was addressed; how the problem was approached; what major results were found; what major conclusions were drawn. And it should do this in a single paragraph! A successful abstract must present a complete and accurate summary of your work and be fully self-contained, making perfect sense to someone who has not read any other part of your report, as in the example below. Abstracts are typically written in the passive voice.

This study was undertaken to determine the wave-lengths of light that are most effective in promoting photosynthesis in the aquatic plant Elodea canadensis, since some wavelengths are generally more effective than others. The rate of photosynthesis was determined at 25°C, using wavelengths of 400, 450, 500, 550, 600, 650, and 700 nm and measuring the rate of oxygen production for 1 h periods at each wavelength. Oxygen production was estimated from the rate of bubble-production by the submerged plant. We tested four plants at each wavelength. The rate of oxygen production at

450 nm (approximately 2.5 ml O_2/mg wet weight of plant/h) was nearly 1.5 × greater than that at any other wavelength tested, suggesting that the light of this wavelength (blue) is most readily absorbed by the chlorophyll pigments. In contrast, light of 550 nm (green) produced no detectable photosynthesis, suggesting that light of this wavelength is reflected rather than absorbed by the chlorophyll.

THE ACKNOWLEDGMENTS SECTION

Most biologists are aided by colleagues in various aspects of their research. It is customary to thank those helpful people in the penultimate section of theses, scientific papers and final-year undergraduate research project reports, but not normally in ordinary undergraduate lab reports. Here is an example that might be found in a typical PhD thesis.

> I am greatly indebted to my supervisor, Dr R. J. Littleton, for his patience, encouragement and guidance throughout this work. At Imperial College, I thank Dr C. Peters and other staff of the Computer Centre for their help with the programming, and in the Biology Department, Mrs R. Chu for help with electron microscopy and Mr J. Dawes for advice on autoradiography. I thank Dr V. Davies of Bristol University for discussions on the population genetics aspects of this research. I am grateful to the SERC for a research studentship.

As in the above example, you must include the last names of the people you are acknowledging and indicate the specific assistance received from each person named.

THE LITERATURE CITED SECTION

In this final section, you present the complete citations for all the factual material you have referred to in your report. This enables the reader (including you at a later date) to obtain quickly the sources you have used. The reader can then obtain additional information about a particular topic and can verify what you have written, as your interpretation or recollection of what you read may be wrong. Proper referencing is even more crucial in theses and scientific publications. Misstatements of fact are readily propagated in the literature by others; the Literature Cited section enables the reader to verify the factual statements made.

Listing the References

Ideally, include only those references that you have actually read. Include all the references that you mentioned in your report or paper. Unless told otherwise, list references in alphabetical order according to the last name of the first author of each publication. If you cite several papers written by the same author, list them chronologically. If one author has published two papers in the same year, list them as, for example, Harris, C. L., 1990a and Harris, C. L., 1990b. Each citation must include the names of all authors, the year of publication, and the full title of the paper, article, or book. In addition, when citing books you must report the publisher, the place of publication and the pages referred to. When citing journal articles, include the name and volume number of the journal and the inclusive page numbers.

Unfortunately, formats differ from journal to journal, so check the format in the journal you are considering. A few rules apply to most journals:

1. Give only the last names of authors; initials are used for first and other names.
2. Latin names, including species names, are underlined to indicate italics.
3. Titles of journal articles are not enclosed within quotation marks.
4. Journal names are usually abbreviated. In particular, the word *Journal* is abbreviated as *J.*, and words ending in *-ology* are usually abbreviated as *-ol.* The *Journal of Zoology* thus becomes *J. Zool.* Do not abbreviate the names of journals whose titles are single words (for example, *Science* or *Nature*). Acceptable abbreviations for the titles of journals can usually be found within the journals themselves.

The most important rule for the Literature Cited section is to provide all the information required, in a consistent manner. When preparing a paper for publication, you should rigidly follow the format used by the journal to which your typescript will be submitted. Note for example whether 'and' or '&' is used, or 'ed.' or 'editor', the punctuation, and whether the volume number is in normal print, italicised or in bold print.

The following examples should be helpful in preparing your Literature Cited section. Note that the last names of all authors of a paper are included, even though the names of only one or at most two authors (*e.g.* Bayne *et al.*, 1976; Eyster and Morse, 1984) are cited in the text of the report. The format for books differs from that used for citing research papers.

Listing Journal References

Bayne, B. L. 1972. Some effects of stress in the adult on the larval development of *Mytilus edulis. Nature* (London) 237: 459.

Carlton, J. T., G. J. Vermeij, D. R. Lindberg, D. A. Carlton and E. C. Dudley. 1991. The first historical extinction of a marine invertebrate in an ocean basin: the demise of the eelgrass limpet *Lottia alveus. Biol. Bull.* 180: 72-80.

Hilbish, T. J. and K. M. Zimmerman. 1988. Genetic and nutritional control of the gametogenic cycle in *Mytilus edulis. Mar. Biol.* 98: 223-228.

Listing Book References

Wessells, N. K. and J. L. Hopson. 1988. *Biology*. Random House, Inc., NY, pp. 374-379.

Listing an Article From a Book

Toole, B. P. 1981. Glycosaminoglycans in morphogenesis. In: *Cell Biology of Extracellular Matrix* (E. D. Hay, editor), Plenum Press, NY, pp. 259-294.

Listing a Laboratory Manual or Handout

Biology 13 Laboratory Manual. 1991. Exercise in Enzyme Kinetics, pp. 16-23. Swarthmore College, PA.

Rogers, C. A. 1991. Principles of physiology, using insects as models. II. Excretion of organic compounds by Malphighian tubules. 2nd Year Animal Physiology Schedule, Leicester University, UK.

A sample Literature Cited section follows, with items arranged alphabetically and chronologically. Your lecturer may specify a different format for this section of your report, so check if you are uncertain.

Literature Cited

Bayne, B. L. 1972. Some effects of stress in the adult on the larval development of Mytilus edulis. Nature (London) 237: 459.

Bayne, B. L., D. R. Livingstone, M. N. Moore and J. Widdows. 1976. A cytochemical and biochemical index of stress in Mytilus edulis L. Mar. Poll. Bull. 7: 221-224.

Biology 13 Laboratory Manual. 1991. Exercise in Enzyme Kinetics, pp. 16-23. Swarthmore College, PA.

Carlton, J. T., G. J. Vermeij, D. R. Lindberg, D. A. Carlton and E. C. Dudley. 1991. The first historical extinction of a marine invertebrate in an ocean basin: the demise of the eelgrass limpet Lottia alveus. Biol. Bull. 180: 72-80.

Eyster, L. S. and M. P. Morse. 1984. Early shell formation during molluscan embryogenesis, with new studies on the surf clam, Spisula solidissima. Amer. Zool. 24: 871-882.

Hilbish, T. J. and K. M. Zimmerman. 1988. Genetic and nutritional control of the gametogenic cycle in Mytilus edulis. Mar. Biol. 98: 223-228.

Lima, G. M. and R. A. Lutz. 1990. The relationship of larval shell morphology to mode of development in marine prosobranch gastropods. J. mar. biol. Ass. U.K. 70: 611-637.

Toole, B. P. 1981. Glycosaminoglycans in morphogenesis. In: Cell Biology of Extracellular Matrix (E. D. Hay, editor), Plenum Press, NY, pp. 259-294.

Wessells, N. K. and J. L. Hopson. 1988. Biology. Random House, Inc., NY, pp. 374-379.

STATISTICAL ANALYSIS

This section is no alternative to a course in biostatistics. Here we explain why and how statistics are used in biology, how the results of the analyses should be incorporated into the lab report, and how to write about your data if you don't analyse them statistically.

What You Need to Know about Tomatoes, Coins, and Random Events

Variability is a fact of biological life: student performance on any particular examination varies among individuals; the growth rate of tomato plants varies among seedlings; the respiration rate of mice held under a given set of environmental conditions varies among individuals; and the amount of time a lion spends feeding varies from day to day and from lion to lion. Some of the variability in our data reflects unavoidable imprecision in making measurements. If you measure the length of a single bone 25 times to the nearest mm, for example, you will probably not end up with 25 identical measurements. But most of the variability you record in a study reflects real biological differences among the individuals in the sample population. Variability, whether in the responses you measure in an experiment or in the distribution of individuals in the field, is no cause for embarrassment or dismay, but it cannot be ignored in presenting your results or in interpreting them.

Suppose we plant two groups of 30 tomato seeds on day 0 of an experiment, and the individuals in group *A* receive distilled water while those in group *B* receive water plus a nutrient supplement. Both groups of seedlings are held at the same temperature, are given the same volume of water daily, and receive 12 hours of light and 12 hours of darkness each day for 10 days. Twenty-six of the seeds sprout under the group *A* treatment and 23 of the seeds sprout under the group *B* treatment. At the end of 10 days, the height of each seedling is measured to the nearest 0.1 cm, and the data are recorded on the data sheet as shown in Figure 29. Note that the units (cm; sample size) are clearly indicated on the data sheet, as is the nature of the measurements being recorded (height after 10 days). The number of samples taken, or of measurements made, is always represented by the symbol *N*.

The question is: did the mineral supplement make a difference in the height of seedlings by day 10 after planting? If all the group *A* individuals had been 2.0 cm tall and all the group *B* individuals had been 2.4 cm tall, we would conclude that growth rates were increased by adding nutrients to the water. If each group *A* individual had been 2.3 cm tall and each group *B* individual had been 2.4 cm tall, we might again suggest that the nutrient supplement improved the growth rates of the seedlings. In the present case, however, there is considerable variability in the heights of the seedlings in each of the two treatments, and the difference in the average height of the two populations is not large with respect to the amount of variation found within each treatment. The heights of group *A* seedlings differ by as much as 1.0 cm (2.8 cm–1.8 cm) and the heights of group *B* seedlings differ by as much as 0.8 cm (2.8 cm–2.0 cm), whereas the average height differences between the two groups of seedlings is only 0.1 cm (2.4 cm–2.3 cm).

The average seedling height in the two populations is certainly different, but does that difference of 0.1 cm in average height reflect a real biological effect of the nutrient supplement, or have we simply not planted enough seeds to be able to see past the variability inherent in individual growth rates? If we had planted only one seed in each group, the two seedlings might have both ended up at 2.6 cm; some seedlings reached this height in both treatment groups, as seen in Figure 29 (next page). On the other hand, the one seed planted in group *A* might have been the one that grew to 2.8 cm, and the one seed planted in group *B* might have been one of the seeds that grew only to 2.2 cm. Or it might have turned out the other way around, with the tallest seedling appearing in group *B*. Clearly, a sample size of one individual in each treatment would have been inadequate to evaluate our hypothesis conclusively. Perhaps 30

Group A seedlings: water only
(height, in cm, after 10 days)

2.1 cm, 2.1, 2.0, 2.8, 2.7, 2.4, 2.3, 2.6, 2.6, 2.5, 2.0, 2.1, 2.8,
2.0, 1.9, 2.8, 2.0, 2.2, 2.6, 1.8, 2.0, 2.2, 2.5, 2.4, 2.3, 2.1

Average = 2.3 cm; N = 26 measurments

Group B seedlings: water plus nutrients
(height, in cm, after 10 days)

2.6 cm, 2.1, 2.0, 2.4, 2.8, 2.6, 2.6, 2.2, 2.7, 2.4, 2.4, 2.3,
2.2, 2.4, 2.6, 2.4, 2.2, 2.4, 2.8, 2.6, 2.5, 2.6, 2.4

Average = 2.4 cm; N = 23 measurments

FIGURE 29.

Data sheet with measurements recorded.

seeds per sample is also inadequate. If we had planted 1,000 seeds, or 10,000 seeds in each group, the differences between the two treatments might have been even less than 0.1 cm . . . or the differences might have been substantially greater than 0.1 cm. If only we had planted more seeds, we might have more confidence in our results. If only we had measured 100,000 individuals, or one million individuals, or . . . .

But wishful thinking has little place in biology; we have only the data before us and they must be considered as they stand. Is the difference between an average height of 2.4 cm for the group *A* seedlings and 2.3 cm for the group *B* seedlings a real difference? That is, is the difference statistically significant?

As another example, suppose we have mated red-eyed fruit flies with brown-eyed fruit flies and, from knowledge of the parentage of these flies, we expect the second generation offspring to have red or brown eyes in the ratio of 3:1. Suppose we actually count 570 red-eyed offspring and 203 brown-eyed offspring, so that slightly fewer than three quarters of the offspring have red eyes. Do we conclude that our expectations have been met, or that they have not been met? Is a ratio of 2.8:1 close enough to our expected ratio of 3:1? Is the observed result statistically equivalent to the expected ratio?

Establishing a Null Hypothesis

We use statistical tests to determine the significance of differences between sampled populations, or differences between results expected and those obtained. To begin, we must precisely define a specific issue (hypothesis) to be tested. The hypothesis to be tested is called the *null hypothesis*, H_0, which always assumes that nothing unusual has happened in the experiment or study. That is, it assumes that the treatment (addition of nutrients, for example) has no effect, or that there are no differences between the results we observed and the results we expected. Examples of typical null hypotheses are:

> H_0: the seedlings in groups A and B do not differ in height (or, the addition of nutrients does not alter growth rates of the seedlings).

H_0: the eye colour of offspring does not differ from the expected ratio of 3:1.

H_0: caterpillars do not show a preference for the diet on which they have been reared.

H_0: average wing lengths do not differ among populations of house flies.

It may seem surprising that the hypothesis to be tested is the one that anticipates no unusual effects; why bother doing the study if we begin by assuming that our treatment will be ineffective, or that wing lengths will not differ from population to population? The null hypothesis is chosen for testing because scientists must be cautious in drawing conclusions. Hypotheses can never be proven; they can only be discredited or supported, and the strongest statistical tests are those that discredit null hypotheses. The cautious approach in testing the effect of a new drug is therefore to assume that it will not cure the targeted ailment. The cautious approach in testing the effects of different diets on the growth rate of a test organism is to assume that all diets will produce equivalent growth. Only if we can discredit the null hypothesis (the hypothesis of no effect) can we tentatively embrace an alternative hypothesis – for example, that a particular drug is effective, or that wing lengths do differ among populations.

Conducting the Analysis and Interpreting the Results

Once we have established our null hypothesis and collected our data, statistical analysis of the data can begin. A large number of statistical tests have been developed, including the familiar Chi-Square test and the Student's *t*-test. The test that should be used to examine any particular set of data will depend on the type and amount of data collected and the nature of the null hypothesis. Once the appropriate test is chosen, by you or your lecturer, the data are manoeuvred through one or more standard formulae to calculate the desired test statistic. This may be a Chi-Square value, a *t*-value or other values associated with different tests; in all cases, the calculated test value will be a single number, such as 0.93 or 129.8. A value close to zero suggests that the data are consistent with the null hypothesis (little deviation from the outcome expected if the null hypothesis is true). A value very different from zero indicates that the null hypothesis may be wrong, since the data obtained are very different from those expected.

Returning to our seedling experiment, we wish to determine if the addition of certain nutrients alters seedling growth rates. The appropriate test here is the *t*-test, so we apply the formula provided in statistics books (see Appendix B, page 251). The value of *t* calculated for data in our tomato seedling experiment is 1.89. This particular value has some probability of turning up if the null hypothesis of no effect is true. If we repeated the experiment exactly as before, using another set of 60 seeds, we would probably obtain a somewhat different result, and the *t*-statistic would have a different value even though the null hypothesis might still be true. If we did five identical experiments, we would probably calculate five different *t*-values. In other words, a statistic may take on a broad range of values even if the null hypothesis is correct, and each of these values has some probability of turning up in any single experiment. But some values are more likely to turn up than others.

Suppose the null hypothesis, that the addition of nutrients does not alter the growth of tomato seedlings over the first 10 days, is actually correct. If we ran our experiment 100 times, we might actually find no measurable difference between the average heights of the seedlings in some of the experiments, so that our calculated *t*-values for these data would be zero. In most of the experiments, we would probably record small differences between the average sizes of seedlings in the two populations (and, for each of these experiments, calculate a *t*-value close to zero), and in a few experiments, purely by random chance, we would probably record large differences (and calculate *t*-values very different from zero, either much larger or much smaller). All these results are possible if we do enough experiments, even though the null hypothesis is correct, simply because the growth of seedlings is variable even under a single set of experimental conditions. The strange result may not come up very often, but there is always some probability that it will occur.

The significance of variability

The important point is that the outcome of a study can vary quite a lot, whether or not the null hypothesis is actually correct. A non-biological example may help to clarify this point. In coin tossing, a fair coin should, on average, produce an equal number of heads and tails. Yet experience tells us that 10 tosses in a row will often produce slightly more of one result than the other. Every now and then, we will actually toss 10 heads in a row, or 10 tails in a row: neither of these results will occur very often, but each will occur eventually if we repeat the experiment enough times.

There is considerable morphological, physiological and behavioural variability in the real world. The only way to know, with certainty, that our one experiment is a true reflection of that world is to measure or count every individual in the population under consideration (for example, plant every tomato seed in the world, and measure every seedling

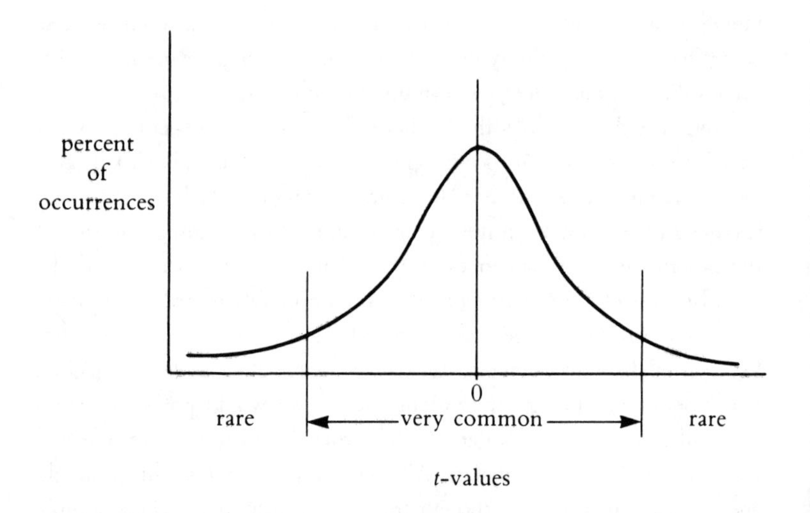

FIGURE 30.

The distribution of *t*-values expected when the null hypothesis (H_0) is true. A wide range of values may occur, but some values will occur more commonly than others. Obtaining a common value for t causes us to accept H_0. Obtaining a rare value for t causes us to doubt the validity of H_0.

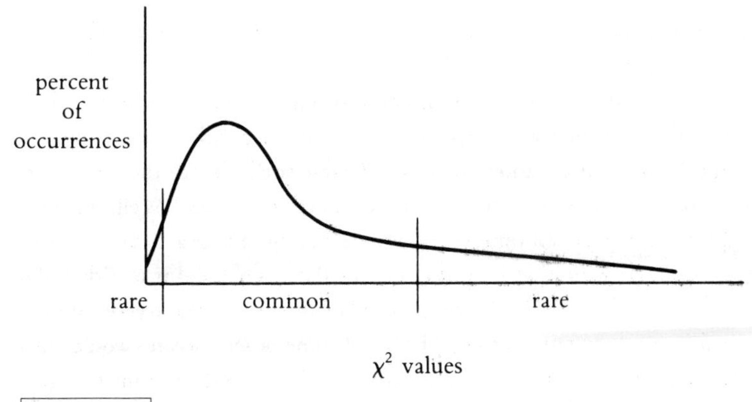

FIGURE 31.

The distribution of χ^2 values expected when H_0 is true. A wide range of values may occur, but some values will occur more often than others. The rarer the value obtained, the less confidence we can have in the validity of H_0.

after 10 days) or conduct an infinite number of experiments. This is not a practical solution to the problem. The next best alternative is to use statistical analysis. Statistics cannot tell us whether we have revealed THE TRUTH, but they can indicate just how convincing or just how strange our results are.

The numerical value of any calculated test statistic has some probability of turning up when the null hypothesis is true. For example, values of *t* are distributed as in Figure 30, and values of Chi-Square (χ^2) are distributed as in Figure 31. If the null hypothesis is correct, values of each statistic will usually fall within a certain range, as indicated; these values will have the greatest probability of turning up in any individual experiment. If the *t*-value calculated for our experiment falls within the range indicated as 'very common,' we are probably safe in accepting the null hypothesis; at least we have no reason to disbelieve it. However, even if the null hypothesis is correct, very unusual values of *t* or of χ^2 will occasionally occur. We are, after all, randomly picking only a few seeds to plant, out of a bag that may contain many thousands of seeds; it could be just our luck to have picked only those seeds that are most unlike the average seed.

If we calculate a very unusual (very high or very low) value for *t* using the data from our experiment, how can we decide to reject the null hypothesis when we know there is still some small chance that H_0 is correct and that we have simply witnessed a very rare event? Well, we must admit that we are not omniscient and be willing to take a certain amount of risk in drawing conclusions from our data; the amount of risk taken can be specified. Typically, researchers assume that if their very unusual (that is, rarely obtained) value of *t* (or of some other statistic) would turn up fewer than 5 times in 100 repetitions of the same experiment when the null hypothesis was true, then this strange value of *t* is a strong argument against the null hypothesis being correct; H_0 is then tentatively rejected. That is, the large calculated value of *t* would be so rarely encountered if the null hypothesis were true that the null hypothesis is *probably* wrong; but there is still the 5 in 100 chance that the null hypothesis is correct and that the researchers, through random chance, happened upon atypical results in their experiment. Tossing 10 heads in a row using a fair coin won't happen very often, but it *will* happen. Tossing 100 heads in a row is an even rarer event, but it could happen. If you conducted only one tossing experiment of 100 flips and tossed only heads, you could tentatively reject the null hypothesis that the coin is fair. But you wouldn't know with certainty that you were correct.

It is, we hope, becoming clear why experiments must be repeated many times before the results become convincing. Such is the challenge of doing biology. Only when large values of a test statistic appear many times can we become fully confident that the null hypothesis deserves to be rejected. Only when low values appear many times can we become confident that the null hypothesis is most likely correct.

Having calculated the test statistic, one consults the appropriate tables for that statistic in a book to determine P, the probability of getting purely by chance a deviation (from the expectations of the null hypothesis) as big as the observed deviation. Check in your statistics book how to calculate the number of *degrees of freedom* for your particular test, then look up the P value corresponding to the value for your test statistic for that number of degrees of freedom.

An Example of a Statistical Test: the Use of Chi-Squared

With χ^2, you need to decide whether to use a goodness-of-fit χ^2, for testing an observed ratio for its fit to an expected ratio, *e.g.* whether a population with 40 males to 30 females fits an expected 1:1 sex ratio, or whether to use a homogeneity χ^2 test. In the latter, one tests whether two sets of observed values are homogeneous (not significantly different), without having an expected ratio. For example, is the sex ratio the same in two populations, one with 40 males and 30 females, and one with 50 males and 30 females?

It is essential to use original numbers in χ^2 tests, never percentages, frequencies or fractions, because the sample size is excluded from such data reductions. Suppose you expected a 1:1 sex ratio and have three populations, all with 40 per cent males. Their respective numbers of males and females are: (1), 2, 3; (2), 20, 30; (3), 20,000, 30,000. With only five individuals in population 1, the closest possible fit to a 1:1 ratio is 2, 3, so the difference cannot be significant. In population 2, there is some deviation from the expected 25, 25 distribution, and in population 3 the deviation from 25,000, 25,000 is enormous and most unlikely to be by chance. Therefore one cannot test the fit by testing 40 per cent males observed against 50 per cent males expected: one must test actual numbers, such as 20:30 observed against 25:25 expected.

As a worked example, consider figures for spontaneous mutations in the evolution of haemoglobins and myoglobins: 292 transition mutations (purines changing to other purines, pyrimidines changing to other pyrimidines) and 548 transversion mutations (purine to pyrimidine change, pyrimidine to purine change) were observed. Each base pair in DNA has one possible transition and two possible transversions, so if the two types of change occur at random, one expects a 1:2 ratio for transitions: transversions. With a total of 840 observations, and a 1:2 (*i.e.* $^1/_3$ to $^2/_3$) expected ratio, we expect $^1/_3 \times 840 = 280$ transitions and $^2/_3 \times 840 = 560$ transversions if these replacements occur at random.

$$\chi^2 = \sum \frac{(\text{observed} - \text{expected})^2}{\text{expected}} = \frac{(292 - 280)^2}{280} + \frac{(548 - 560)^2}{560} = 0.51 + 0.26 = 0.77.$$

We now need to work out the number of degrees of freedom, which for χ^2 is normally one fewer than the number of classes. Here we have two classes, transitions and transversions, and their total number, 840. Within the given total, if we fix the number in one class, say at 292 transitions, then the other class cannot vary, as it must have 548 transversions when the total is 840. So we look up $\chi^2 = 0.77$ for one degree of freedom, finding that P lies between 0.50 and 0.30, so we would expect a deviation as big as the one observed here to occur purely by chance in 30–50 per cent of all experiments. $P > 0.05$, so the observed difference is not significant at the 5 per cent level. We can accept the null hypothesis that base pair replacements during DNA evolution of these globin molecules occurred at random, not in a biased fashion.

Do remember to calculate *(observed – expected)²/expected* separately for all classes. So if you were testing a fit to a 9:3:3:1 ratio, you would have four classes, four *(observed – expected)²/expected* calculations to do, and three degrees of freedom (*i.e.* one fewer than the number of classes). Showing that your data fit a particular expected ratio does not prove that that ratio is the correct one. For example, 67:55 is an acceptable fit to both a 1:1 and a 9:7 ratio.

Summary

In summary, in performing a statistical test you first decide on a reasonable degree of risk (usually 1 per cent or 5 per cent) of incorrectly rejecting the null hypothesis, perform the study, put the data into the appropriate formula to calculate the value of the appropriate statistic, and end up with a single number. You then look in the appropriate statistical table to see where this number is within the range of values expected when the null hypothesis is correct. If your number lies within the expected range of values, your data support the null hypothesis. If your number lies outside the range of commonly expected values, your data do not support H_0; they support the alternative hypothesis. But remember, there is always some small chance that your null hypothesis is correct and that you are making the wrong decision by rejecting H_0 (false negatives). Similarly, when your number lies within the range of commonly recorded values, there is always some chance your null hypothesis is actually incorrect and that you are making the wrong decision by accepting it (false positives). For this reason, we biologists never seek to *prove* any particular hypothesis; we can only accumulate data that either favour or argue against the null hypothesis.

Incorporating Statistics into Your Lab Report

If you conduct no statistical analyses, be especially cautious when making statements about your data. Your data may support one hypothesis more than another, but they cannot prove that any hypothesis is true. In addition, if you conduct no statistical analyses, you cannot say that differences between groups of measurements are significant or not significant. *Significance* implies subjecting data to rigorous statistical testing. It is perfectly fair to write that 'seedlings treated with nutrients appeared to grow at slightly faster rates than those treated with distilled water' and refer the reader to the appropriate table or figure, but you cannot say that seedlings in one treatment grew significantly faster than those in the second treatment.

Include any statistical analyses of your data in your Results section. Use the results of the analyses to support any major trends that you see in your data, as in the following two examples.

Example 1

> For thirty caterpillars reared on the mustard-flavoured diet and subsequently given a choice of foods, the caterpillars showed a statistically significant preference for the mustard diet ($\chi^2 = 17.3$; $P < 0.05$). For thirty caterpillars reared on the quinine-flavoured diet, there was no influence of previous experience on the choice of food ($\chi^2 = 0.12$; $P > 0.10$).

In this example, H_0 is that prior experience will not influence the subsequent choice of food by caterpillars; '$P < 0.05$' means that if the experiment were repeated 100 times and H_0 were true, such a high value for χ^2 would be expected to occur in fewer than 5 of those 100 studies. In other words, the probability of making the mistake of rejecting H_0, when it is in fact true, is less than 5 per cent. You can therefore feel reasonably safe in rejecting H_0 in favour of the alternative: that prior experience does influence subsequent food selection for caterpillars reared on the mustard diet.

Different results were obtained, however, for the caterpillars reared on the quinine-flavoured diet. In example 1, sentence two, 'P > 0.10' means that if the experiment were repeated 100 times and H_0 were true, you would expect to calculate such a small value of χ^2 in at least 10 of the 100 trials. In other words, the probability of getting this χ^2 value with H_0 true is rather high; certainly the χ^2 value is not unusual enough for you to mistrust H_0 and run the risk of rejecting the null hypothesis when it is in fact true.

Example 2

> Over the first 10 days of observation, growth of seedlings receiving the nutrient supplement was not significantly faster than the growth of the seedlings receiving only water ($\underline{t}$ = 1.89; P > 0.10).

In this second example, the null hypothesis (H_0) states that the nutrient supplement does not influence plant growth; 'P > 0.10' again means that if the experiment were repeated 100 times and H_0 were true, you would expect to calculate such a low value of *t* in more than 10 of the 100 trials. You have obtained a value of *t* that would be common if H_0 were true and so have no reason to reject H_0. It is possible that H_0 is actually false and the nutrients really do promote seedling growth, and that you just happened upon an unusual set of samples that gave a misleadingly small *t*-value. If such is the case, repeating the experiment should produce different results and larger *t*-values. But with only the data before you, you cannot reject H_0.

Note that you need say little about the statistics themselves when writing your report. You would simply state, in your Materials and Methods section, that the data were analysed by Chi-square or some other test, and then, in the Results section, include a few test statistics to back up your interpretations of data as indicated above; the few sentences of Examples 1 and 2 tell the complete story.

Describing the Results of Statistical Tests

If you have determined from your test statistic that the probability value is 0.20, P = 0.20, you should describe this as showing that the probability of getting your observed deviation from the expectations of the null hypothesis, purely by chance, is 0.20 (as a fraction), or 20 per cent. Therefore the deviation is acceptable as being due to chance, so the null hypothesis (state again what this is) is supported, not the alternative hypothesis. For example, the observed ratio fits the expected one, or the two populations are not significantly different.

Here, from students, are some poor descriptions of statistical results:

> Therefore the above results occurred by chance, not by an external factor.

This is wrong because the deviation, not the results, occurred by chance.

> There is a 20–30 per cent chance that the fit occurred by chance.

No; the deviation, not the fit, occurred by chance.

> The ratio can be deemed to be significant.

No; because the deviation is *not* significant, the observed ratio fits the expected ratio.

PREPARING THESES

Many aspects of preparing theses have already been covered in this chapter, as they are similar to those for lab reports. A thesis is longer and more formal, however, so do check the appropriate regulations, such as maximum length and style of binding. A thesis is an organised account of the research you have done. It must have an up-to-date literature survey and must make clear the relations between your work and previous work. Towards the end of the Discussion, you should give some carefully thought-out suggestions for future research, building on your findings. You never have enough time to do all the experiments you want to, so suggest those which you would like to have done had time and resources permitted.

During your research and its writing up, keep in close touch with your supervisor, and seek advice from anyone who can help with materials, advice, equipment, techniques, knowledge or suggestions. Acknowledge any such help in the Acknowledgments section (see page 110). Do get a draft of your thesis, or each chapter in turn, looked at by someone else who can comment on what is not clear to them and on any scientific or English errors.

A thesis usually takes *at least* twice as long to write as people expect, so start as early as possible, perhaps on the literature survey and Materials and Methods. A good lab notebook for ideas, reasons, methods, results and analyses is essential, with more than one copy of each part in case of loss, theft, fire, etc. Deposit a copy of key results with your supervisor, and give him or her a copy of the final thesis to keep, as well as draft versions to comment on.

There is no need to present your ideas or results in a chronological order, if some other order is more logical or is clearer. In the thesis, the usual order of parts is: Title Page, Abstract, Table of Contents, Acknowledgments, Introduction (including literature survey), Materials and Methods, Results, Discussion (including suggestions for future work), Appendix (this is optional, and is sometimes used for detailed data not immediately used in the text, long series of equations or detailed statistical analyses), References. There may also be a Conclusions section after the Discussion, or a Summary section, summarising the whole thesis. A detailed Table of Contents is very useful to you and your readers for finding specific pieces of information. In unusual cases where several very diverse topics have been studied, each topic may have its own Title, Introduction, Materials and Methods, Results, and Discussion, but there should still be a general overall introduction and a general discussion to bring all the parts together.

If you use a typist, give her or him precise instructions about indentation, margins, spacing, placing of tables, graphs and drawings, layout of headings, captions, tables, underlining, italics, type of printer to use, etc. Allow time for corrections, and discuss costs in advance.

Consult your supervisor about style. Most prefer an impersonal style, not using 'I' or 'we' unless strictly necessary, which it is for comments such as: '*I* suggest. . .'. A long research project involves a great deal of hard work; make sure your thesis does it justice.

4. Writing Essays

The term 'essay' will here include ordinary undergraduate essays, perhaps for tutorials, and long essays involving a lot of literature-searching, and which may be called 'reviews' or 'dissertations'. Each of these tasks requires you to present critical evaluations of what you have read. In preparing an essay, you synthesise information, explore relationships, analyse, compare, contrast, evaluate, and organise your own arguments clearly, logically, and persuasively, leading up to an assessment of your own. A good essay is a creative work; you must interpret thoughtfully what you have read and come up with something that goes beyond what is presented in any single book or article consulted.

At first, undergraduates mainly use textbooks and their lecture notes, but later they make more use of primary scientific literature – that is, the original research papers published in such scientific journals as *Biological Bulletin*, *Cell*, *Developmental Biology*, *Ecology*, and *Journal of Comparative Biochemistry and Physiology*. Textbooks and review articles (such as those in *Scientific American* or *Quarterly Review of Biology*) are secondary literature, giving someone else's interpretation and evaluation of the primary literature. The literature and how to search it were covered in Chapter 2.

THE VALUE OF ESSAYS

Every time your class is asked to write an essay, your lecturer is committing himself or herself to many hours of reading and marking. There must be a good reason to require such assignments; most staff are neither masochists nor sadists.

Writing essays benefits you in several important ways. For one thing, it allows you to teach yourself something relevant to the course you are taking. The ability to self-teach is essential for success in graduate programmes and in almost any career. Additionally, you gain experience in reading scientific literature. Textbooks and many lectures present you with facts and interpretations. By reading the papers upon which these facts and interpretations are based, you come face-to-face with the sorts of data, and interpretations of data, that the so-called facts of biology are based on, so you gain insight into the true nature of scientific enquiry. Preparing thoughtful essays will help you move away from the unscientific, blind acceptance of stated facts toward the scientific, critical evaluation of data. These assignments are also superb exercises in the logical organisation, effective presentation and discussion of information. How fortunate you are that your lecturer cares enough about your future to give such assignments!

There is another reason why lecturers ask their students to prepare essays. One can summarise a dozen papers without understanding the contents of any of them. This is the book report format, in which the writer merely presents facts uncritically: this happened, that happened; the authors suggested this; the authors found that. By writing an essay rather than a book report, you can show your lecturer that you understand what you have read, that you have really learned something.

GETTING STARTED

Your lecturer will usually give you the topic or title for your essay. If you are given a choice, choose a subject that interests you. If you have a choice of broad subjects and start reading on one topic, you will usually find that you must narrow your focus to a particular area within that broad topic, because you encounter an unmanageable number of references. You cannot, for instance, write about the entire field of primate behaviour because the field has many different facets, each associated with a large and growing amount of literature. In such a case, you will find a smaller topic, such as the social significance of primate grooming behaviour, to be more appropriate. As you continue your literature search, you may even find it necessary to restrict your attention to a few primate species. Alternatively, you may find that the topic originally selected is too narrow and that you cannot find enough information on which to write a substantial essay. You must then broaden your topic or switch topics entirely.

RESEARCHING YOUR TOPIC

As discussed in Chapter 2, begin by carefully reading the appropriate section of your lecture notes and textbook to get an overview of the general subject area, and then perhaps consult at least one more specialised book before tackling more specialised literature. For some first-year undergraduate essays, much of the needed information may be in your lecture notes, with no need for you to study original papers. For more advanced essays, you are now ready to locate and read research reports on your topic, following the advice in Chapter 2. The goal is to select a number of interrelated papers and to read these with considerable care, not simply to accumulate a huge number of references that then receive cursory attention. Continually ask yourself while taking notes, 'Why am I writing this down? What is especially relevant and interesting about this particular piece of information? Can I see any relationship between this information and what I have already written or learned?'

WRITING THE ESSAY

Begin by reading all your notes, preferably without pen or pencil in hand. Having read your notes to get an overview, reread them with the intention of sorting your ideas into categories. Notes taken on index cards are particularly easy to sort, provided that you have not written many different ideas on a single card; one idea per card is a good rule to follow. To arrange notes written on full-sized sheets of paper, some people annotate the notes with pens of different colours or using a variety of symbols, with each colour or symbol representing a particular aspect of the topic. Other people use scissors to snip out sections of the notes (which must be on one side of the paper only) and then group the resulting scraps of paper into piles of related ideas.

At this point you must eliminate those notes that are irrelevant to the specific topic you have finally decided to write about. Some of your earlier notes are likely to be irrelevant if they were taken before you had developed a firm focus. Don't let irrelevant facts find their way into your essay, however interesting they are. Decide next how best to arrange your categorised notes so that your essay progresses towards a conclusion. Again, ask yourself whether a particular section of your notes seems interesting to you, and why it does, and look for connections among the various items as you sort.

The direction your essay will take should be clearly and specifically indicated in the opening paragraph, as in the following example written by Student *A*:

> Most lamellibranch bivalves are sedentary, living either in soft-substrate burrows (e.g., soft-shell clams, Mya arenaria) or attached to hard substrate (e.g., the blue mussel, Mytilus edulis) (Barnes, 1980). However, individuals of a few bivalve species live on the surface of substrates, unattached, and are capable of locomoting through the water. One such species is the scallop Pecten maximus. In this essay, I will explore the morphological and physiological adaptations that make swimming possible in P. maximus, and will consider some of the evolutionary pressures that might have selected for these adaptations.

The nature of the problem being addressed is clearly indicated in this first paragraph and Student *A* tells us clearly why the problem is of interest: (1) the typical bivalve doesn't move and certainly doesn't swim; (2) a few bivalves can swim; (3) so what is there about these exceptional species that enables them to do what other species can't; (4) and why might this swimming ability have evolved? Use of the pronoun *I* is now acceptable in some scientific writing, although some staff – and some scientific journals – prefer an impersonal style.

In contrast to the previous example, consider the following weaker (though not altogether bad) first paragraph written by Student *B* on the same subject.

> Most bivalved molluscs either burrow into, or attach themselves to, a substrate. In a few species, however, the individuals lie on the substrate unattached and are able to swim by expelling water from their mantle cavities. One such lamellibranch is the scallop Pecten maximus. The feature that allows bivalves like P. maximus to swim is a special formation of the shell valves on their dorsal sides. This formation and its function will be described.

In this example, the second sentence weakens the opening paragraph considerably by prematurely referring to the mechanism of swimming. The main function of the sentence should be to emphasise that some species are not sedentary; the reader, not yet in a position to understand the mechanism of swimming, becomes a bit baffled. The penultimate sentence of the paragraph ('The feature that allows . . .') also hinders the flow of the argument. This sentence summarises the essay before it has even been launched, and again, the reader is not yet in a position to appreciate the information presented; what is this 'special formation' and how does it have anything to do with swimming? The first paragraph should be an introduction, not a summary.

The last sentence of Student *B*'s paragraph does clearly state the objective of the paper, but the reader must ask, 'Towards what end?' The author has set the stage for a book report, not an essay. Reread the paragraph written by Student *A* and notice how the same information has been used much more effectively, introducing a thoughtful essay rather than a book report. Student *A*'s first paragraph was written with a clear sense of purpose, with each sentence carrying the reader forward to the final statement of intent. You might guess, correctly, from Student *B*'s first paragraph that the rest of the essay was somewhat unfocused and rambling. In contrast, Student *A*'s first paragraph clearly signals that what follows will be well-focused and tightly organised. Get your essays off to an equally strong start.

Another example of a typical, but not especially effective, first paragraph is helpful here:

> The crustaceans have an exceptional capability for changing

> the intensity and pattern of their colouring (Russell-Hunter, 1979). Many species seem able to change their colour at will. The cells responsible for producing the characteristic colour changes of crustaceans are the chromatophores. The function of these cells will be discussed in this essay.

The author is off to a strong start with the first sentence. The second sentence, however, begins by repeating information already given in the first sentence (crustaceans can change colour), and ends by saying nothing at all (what does 'at will' mean for a crustacean?). The last sentence sets up a book report. Why will the function of these cells be discussed? More to the point, why should the reader be interested in such a discussion? The reader will be more readily drawn into your intellectual net if you indicate not only where you are heading but also why you are undertaking the journey. Your first paragraph must state clearly what you are setting out to accomplish and why. Every paragraph that follows should advance your argument clearly and logically towards the stated goal.

Use your information and ideas to build an argument, to develop a point, to synthesise. In essays requiring you to read original papers, avoid the tendency simply to summarise papers one by one: they did this, then they did that, and then they suggested the following explanation. Instead, set out to compare, to contrast, to illustrate, to discuss. Back up your statements with examples drawn from the literature, referring to papers that support your statement, as in the following examples:

> The ability of an organism to recognise 'self' from 'non-self' is found in both vertebrates and invertebrates. Even the most primitive invertebrates show some form of this immune response. For example, Wilson (1907) found that disassociated cells from two different species of sponge would regroup according to species; cells of one species never aggregated with those of the second species.

> Schistosomiasis is one of the most serious parasitic diseases of mankind, afflicting hundreds of millions of people

and causing hundreds of millions of dollars in economic losses yearly through livestock infestations (Noble and Noble, 1982).

Don't simply state that a particular experiment supports some particular hypothesis; briefly describe relevant parts of the experiment and explain how the results relate to the hypothesis under question, as demonstrated in the following two examples.

Foreign organisms or particles that are too large to be ingested by a single leukocyte are often isolated by encapsulation, with the encapsulation response demonstrating clear species-specificity. For example, Cheng and Galloway (1970) inserted pieces of tissue taken from several species into an incision made in the body wall of the gastropod Helisoma duryi. Tissue transplanted from other species was completely encapsulated within 48 hours of the transplant. Tissue obtained from individuals of the same species as the host was also encapsulated, but encapsulation was not completed for at least 192 hours.

Above a certain temperature, further temperature increases often have a depressing effect on larval growth rates (Kingston, 1974; Leighton, 1974). This break point can be very sharply defined. For instance, larvae of the bivalve Cardium glaucum were healthy and grew rapidly at 31°C, grew abnormally and less rapidly at 32–33°C, and grew hardly at all at 34°C (Kingston, 1974).

In all your writing, avoid quotations unless they are absolutely necessary; use your own words whenever possible. At the end of your essay, summarise the problem addressed and the major points you have made, as in the following example:

In conclusion, the basic molluscan plan for respiration that had been successfully adapted to terrestrial life in one group

> of gastropods, the terrestrial pulmonates, has been successfully readapted to life in water by the freshwater pulmonates. Having lost the typical molluscan gills during the evolutionary transition from salt water to land, the freshwater pulmonates have evolved new respiratory mechanisms involving either the storage of an air supply (using the mantle cavity) or a means of extracting oxygen while under water, using a gas bubble or direct cutaneous respiration. Further studies are required for fully understanding the functioning of the gas bubble in pulmonate respiration.

Never introduce any new information in your summary paragraph.

CITING SOURCES

This topic is covered in Chapter 3 (pages 90–4).

CREATING A TITLE

By the time you have finished writing, you should be ready to entitle your creation, unless the title has already been fixed by your tutor. Give the essay a title that is appropriate and interesting, one that conveys significant information about the specific topic of your paper. This is covered in Chapter 3 (pages 108–9).

REVISING

Once you have a working draft of your essay, you must revise it, clarifying your presentation, expunging ambiguity, eliminating excess words and improving the logic and flow of ideas. Always edit for grammar, punctuation and spelling. These topics are considered in detail in Chapters 5 and 6.

5. Improving Punctuation, Word-Choice, Spelling and Grammar

Improving your command of the English language will obviously improve your writing about biology, as well as your job applications, personal letters and creative writing, because it improves your clarity and powers of expression. This chapter gives advice on the most relevant parts of the language.

Unfortunately, tutors often set students a bad example, with errors in handouts, board work, overhead transparencies, draft papers and even in exam papers. One secretary wrote: 'Any senior secretary will tell you about the poor standards of English among academics. We are the ones who try to knock their ungrammatical, misspelled and badly punctuated written work into shape, although we are often impeded by the arrogance of those who cannot accept advice from a non-graduate.' I (BL) greatly admire the clarity of that secretary's English, but have to add that some secretaries' English is less than perfect.

I (BL) found out what areas of students' English most need attention from my study, *A National Survey of UK Undergraduates' Standards of English* (1992, The Queen's English Society, London). In results from 17 UK universities, the following percentages of biology students were recorded as 'poor' at these aspects of English: spelling, 23 per cent poor; punctuation, 31 per cent; grammar, 25 per cent; handwriting, 26 per cent; vocabulary, 24 per cent; clarity of written English, 30 per cent; clarity of spoken English, 23 per cent. Those not rated 'poor' were generally rated 'adequate' rather than 'good'. The percentages of biological departments wanting tuition for their students in various aspects of English, if time and resources permitted, were: written English, 47 per cent; spoken English, 35 per cent; grammar, 83 per cent; spelling, 75 per cent; punctuation, 75 per cent; handwriting, 45 per cent; clarity of written English, 85 per cent; clarity of spoken English, 42 per cent. There is clearly a strong need for biologists to improve their English.

ENGLISH TERMINOLOGY

To discuss how to use, improve or repair a car, it helps to know the basic terminology, such as *engine, wheels, ignition, steering, brakes* and *fuel pump.* Similarly, to improve your English, you need to know some basic words about the language. Pass on to the next section if you know the terms already; consult a textbook on English if you need more detail.

Parts of Speech

Words are classified by function, as follows:

- **Nouns** name things, people, places and ideas. Proper nouns have a capital letter, do not usually have 'the' or 'a' before them, and name such things as individuals, places, days, *e.g. Peter, London, Monday.* Countable nouns name things which can be counted, can be preceded by 'the' or 'a', and usually have singulars and plurals: a *stamen,* four *whales.* In the singular, they usually need a word such as 'a' or 'that' before them, but not in the plural: A *lion* is. . .; *Lions* are. . .; but not: *Lion* is. . . Uncountable nouns do not take numbers and usually have no plural; they may be 'concrete' if touchable: *flour, pepper* (in the sense of the condiment pepper), or abstract: *beauty, intelligence.* Many words are countable in one meaning and uncountable in another: *pepper,* in the sense of a green pepper fruit, is countable, as is *beauty* in the sense of a beautiful person.

- **Pronouns** stand in place of nouns, *e.g. it, they, them, anyone, whom.*

- **Adjectives** qualify (describe or limit) nouns or pronouns: a *red* stain; *two* years; *better* separations; Joan is *brilliant.* Nouns can also be used as adjectives to qualify nouns: *University Technicians' Salary* Report – the first three words are used as adjectives to tell us more about the noun *Report.* If adjectives are used before two consecutive nouns, there can be ambiguity as to which noun the adjective applies to, as in *The green bird's nest,* where one is not sure whether the bird or the nest is green. *The nest of the green bird,* or *The green nest of the bird,* would be clearer.

- **Verbs** are words usually expressing an action, or state of being, telling you about their subject, and having tenses (such as present, past, future): he *draws*; lions *attacked* him; larvae *will become* adults. Transitive verbs need a direct object (*him* in: lions *attacked* him) to undergo the action, as well as a subject to do the action (*lions* in: lions *attacked* him). Intransitive verbs have no direct object: It *hibernated.* Verbs about *being, becoming* and *seeming* ('stative verbs') link their subject with a descriptive complement: Roger *is* a molecular biologist; the students *seemed* bored. A finite verb is a verb in a state limited by its subject in number (singular or plural), and limited by tense and mood, *e.g.* he *says*, as opposed to the infinitive state (*e.g. to say*) or a participle (present participle, *saying*; past participle, *said*) or gerund (a noun formed from the verb by adding *-ing*: His *saying* that was unwise). Knowing what a finite verb is will help you to avoid incomplete sentences, lacking a finite verb, such as: *Estimating molecular weights from using known marker DNA.* That has only two gerunds, with no finite verb. It would be all right as a title or heading, as those do not need to be in complete sentences. Incidentally, headings or titles do not have a full stop at the end, when they are not complete sentences.

- **Adverbs** modify verbs, adjectives and other adverbs, answering questions about how? when? where? how often or how much?: the reaction worked *quickly;* the pupae hatched *late*; I went *there*; he failed *twice.* Many adverbs end in *-ly*, being formed by addition of *-ly* to the adjective: normal+*ly* = *normally*, which helps you to get the correct number of *l*'s in such words. Examples of adverbs modifying adjectives are: He was *very* slow; sloths are *extremely* lazy. Adverbs can modify adverbs: the autoradiographs turned out *really* beautifully; their evolution happened *very* quickly. You need to know the difference between adverbs and adjectives in order to avoid the error of using adjectives for adverbs and vice versa. What is wrong with: *The reaction worked real quick*? The adverb *quickly* is needed, not the adjective *quick*, to qualify the verb *worked*, and the adverb *really* is needed to qualify the adverb *quickly*.

- **Conjunctions** join or show relations between words, phrases and clauses: a rapid *and* complete response; a yellow pigment *or* a green one; he ran *although* she walked. Other conjunctions include *because, but, neither. . .nor.* It helps to know what conjunctions are for when we come to rules for punctuation.

- **Prepositions** are usually positioned before nouns or pronouns to show relationships of position, time, reason, etc.: *on* the bench; *at* work; *until* next year; *because of* his skills.

- **Interjections** are exclamatory words or phrases, such as *ugh! oh dear!*, and are rarely used in written science.

Phrases, Clauses and Sentences

- **Phrases** are groups of words without a finite verb (see definition above) but forming a unit of sense; they can act as nouns, adjectives or adverbs: *an expensive electron microscope* (noun-phrase); the student *having the highest marks* was delighted (adjective-phrase describing *student*) ; give it *to the technician* (adverb-phrase, answering *where* to give it).

- **Clauses** are groups of related words containing a finite verb. A clause may be a one-clause sentence: *We went to lunch,* or part of a compound sentence with more than one main clause: *We went to lunch and returned quickly* (the second clause starts at *and*, with *we* as the understood subject of the verb *returned*). It may be part of a complex sentence, which includes a subordinate clause: *We went to lunch* (end of main clause) *because we were starving* (subordinate clause acting as an adverb for *went*, explaining *why*). Main clauses can usually stand as sentences on their own, but subordinate clauses, acting as adjective-, adverb- or noun-clauses, depend on the main clause to make sense, as in the above example.

- **Sentences** are usually sets of words starting with a capital letter and ending in [.] or [?] or [!], making a complete grammatical structure, although one can get single-word sentences, especially in dialogue when the other words are understood but not included: *Yes.* (*Yes, I will come,* might be the full implied sentence.) As noted above, sentences can be complex or compound, but are called 'simple' if composed of one clause only.

- **Paragraphs** normally develop one topic, consisting of one, or usually more than one, sentence. Try to start your paragraph with a sentence covering the main topic of the paragraph, and with a link to the previous paragraph if possible.

Other Useful Terms

When we come to rules for spelling, we will need to know some other terms.

- **Vowels** are the letters *a, e, i, o, u,* and the letter *y* sometimes acts as a vowel, as in *rhythm.*

- **Consonants** are the 21 non-vowel letters, *b, c, d, f,* etc., including *y*, as in *yet.*

- **Prefixes** are a letter or a group of letters placed at the beginning of the main (base) word to modify its meaning, *e.g. ab-* (*ab* + base word *normal = abnormal*), *dis-, hyper-.*

- **Suffixes** are a letter or a group of letters placed at the end of a word to modify its meaning, *e.g. -ful* (*regret+ful = regretful*), *-ing, -ment.*

- **Syllables** in words are audible units containing a vowel sound. *Big, type* and *through* have one unit of sound each, with one vowel sound, irrespective of the number of written vowels. *Shoulder* has two syllables, which can be represented as *shoul'-der*, and *salary* (*sal'-a-ry*) and *nucleus* (*nuc'-le-us*) have three.

- **Stressed syllables**, indicated in the above examples by ['] at the end of the syllable, are the ones in words of more than one syllable which are stressed in speech, by being pronounced with more emphasis. Read the three examples above out loud. *Shoul'-der* is natural, but *Shoul-der'* seems unnatural. If you are unsure of where syllables end or stresses come, say the word aloud.

- **Long or short vowels**: these terms refer to the length of the vowel sound in speech. In good dictionaries, syllables, stresses and vowel lengths are all indicated in the pronunciation guide to each word. Long vowels are indicated by a bar [¯] over the vowel, and short vowels have no bar, or sometimes are separately indicated by a [˘] over the vowel: *improve* would thus be shown as *im-prōōv'* or *ĭm-prōōv'*.

PUNCTUATING FOR CLARITY AND EASE OF READING

The Use of Different Punctuation Marks

In speech you can use pauses, gestures, facial expressions and changes of pitch and loudness to make your message clear, but in writing you use punctuation marks. Punctuation is a series of marks used to separate words and groups of words, to make the intended meaning clear and easy to follow. It can be used to emphasise certain words and phrases, and to distinguish between major and minor ideas. Skilful punctuation is the key to good sentence construction and therefore to clear expression. It is much more important to understand how punctuation works, by studying examples here and elsewhere, than to memorise long lists of all uses of all punctuation marks. Square brackets [ ] are used to enclose examples of punctuation marks when necessary to avoid confusion in the sentence quoting them.

Bad punctuation can have serious consequences, leading to difficulties in understanding and even to complete misunderstandings. Consider this sentence from a newspaper: *Don't pick up heavy weights like groceries or children with straight legs.* That is hard to make sense of. Better punctuation makes it somewhat clearer: *Don't pick up heavy weights – like groceries or children – with straight legs.* A station was mistakenly demolished in 1984 because a comma was missing from the planning document. The list of items to preserve should have read: *Retain Drem Station, bridge, ...* As the comma after *Station* was missing, the station was demolished and had to be rebuilt, although the bridge was saved. *The Daily Telegraph* reported (6/9/1991) that *The cost of wholesale funds fell to* $10^{1}/_{4}$ *per cent below the present level of base rates. . . .* As the latter level was $10^{1}/_{2}$, the omission of the crucial comma after *per cent* implied that the cost of wholesale funds was $^{1}/_{4}$ per cent instead of $10^{1}/_{4}$ per cent, a vast difference.

The amount of punctuation a sentence requires depends on the complexity of the sentence and of the ideas presented in it, and on the purpose of the writing. In general, the longer and more complex the sentence, the more punctuation is required to make the meaning clear to the reader. If you are unsure of where punctuation is required, try reading the sentence aloud and note where your voice puts in pauses to make the sense clear. Punctuation is often used to separate groups of related words within a sentence, such as clauses and phrases, showing that those words form a sub-unit of sense.

Sentence Beginnings and Endings

The principal unit of writing is the sentence, a group of words (occasionally just one word) which makes sense on its own. Sentences, like this one, start with a capital letter. They end with:

- A full stop [.] for statements: *It is an orchid.*
- A full stop for ordinary requests: *Please pass the pipette.*
- A full stop for mild exclamations: *Well, we have nearly finished the*

experiment.

- A question mark [?] for direct questions: *Are you ready?*
- An exclamation mark [!] for strong exclamations: *What a dreadful mess you've made in this lab!*
- An exclamation mark for strong commands: *Get out of my sight!*

The Full Stop [.]

1. **This is the strongest punctuation mark**, making the most definite pause (in reading aloud or silently) when used at the end of a sentence. As shown above, it is used at the end of sentences unless they are questions, strong exclamations or strong commands.

2. **It is also used to indicate omitted letters in abbreviations**, such as *Mon.* for Monday or *a.m.* for *ante meridiem* (Latin for 'before noon'), and in initials, as in *B.K. Smith.* Common abbreviations, and those of scientific terms and names of organisations, are now usually spelled without full stops, *e.g. Mr* (Mister), *Dr* (doctor), *DNA* (deoxyribonucleic acid), *cm* (centimetre), *UK* (United Kingdom). If an abbreviation ending in a full stop, such as *etc.* (*et cetera,* Latin for 'and so forth'), comes at the end of a sentence, there is no need for another full stop to end the sentence: *I will send you the members' names, addresses, etc.*

3. **Three full stops together (the ellipsis) [. . .] are used to show an unfinished sentence, omission of part of a sentence, or hesitation in speech:** *She would invite him to. . . . No, that was unthinkable.* **Three full stops, with a space between each, are used for an omitted part within a sentence, and four at the end of a sentence.**

The Question Mark [?]

This is used to end a direct question, where an answer is normally expected: *Where do I buy a lab coat?* It is not used for an indirect question (which reports a direct question), to which no answer is expected: *She asked where she could buy a lab coat.*

A question mark is needed when question phrases (question tags) are added to statements: *This is a migratory species, isn't it?*

The Exclamation Mark [!]

This is used after exclamations showing a high degree of surprise (*Fancy meeting you here!*) or strong emotion (*You filthy cheat!*) or special emphasis (*Your theory is so beautiful!*).

It is also used after strong commands or requests, especially where the voice would be raised in speech: *Don't touch that switch!*

Mild requests or commands usually end with a full stop: *Show me your essay, please.*

Using too many exclamation marks weakens their impact. Beware of using an exclamation mark after a number written as figures, as an exclamation mark is a mathematical symbol for 'factorial'.

The Comma [,]

A student wrote: *E. coli has a broad range of habitats – in the gut of some animals in sewage works, etc.* The missing comma after *animals* incorrectly restricted this bacterium to animals in sewage works, so commas do affect the sense, as well as ease of reading.

A comma has many uses, including:

1. **To separate items in a list:** *We collected grasses, sedges, beetles and large fungi.* Commas can also be used to separate a series of phrases or clauses.

2. **To separate two or more adjectives which individually modify a noun:** *He was a small, shy, sickly, red-headed child.* There is no comma after the last adjective, and the commas carry the sense of 'and'. If the last adjective and the noun form a single unit of meaning, there is no need for a comma before the final adjective: *He was a great mathematical genius.*

3. **In pairs, commas are used to separate descriptive phrases or clauses, or less important material, from the main part of the sentence:** *The midwife's car, painted a vivid orange, was parked illegally.* Omitting the first comma would suggest that the car could paint; omitting the

second comma initially suggests that a vivid orange was parked: both commas are needed, operating as a pair. The reader should not have to re-read a sentence to make sense of it.

It is most important to know the difference between phrases or clauses which merely comment, which have a pair of commas separating them from the main part of the sentence, and phrases or clauses which are defining, where a pair of commas would give the wrong meaning. Thus in *The boys, who were fit, enjoyed the race,* 'who were fit' is commenting; this implies that all those particular boys were fit and all enjoyed the race. In *The boys who were fit enjoyed the race,* 'who were fit' is defining: only those boys who were fit enjoyed the race; by implication, those boys who were not fit did not enjoy the race. If only one of these two commas were present, the sentence would be wrong, and the reader would be left unsure whether *who were fit* was commenting or defining. Using a single comma, when either no comma or a pair of commas is required, is a very frequent error in biology and elsewhere.

Note how the presence or absence of one comma can change the meaning:
(i) *She liked Tony, who played football better than John.*
(ii) *She liked Tony, who played football, better than John.*
In (i), *better* refers to *played football,* but in (ii), *better* refers to *liked Tony,* as the words between commas are now a descriptive aside.

Note also the effect of omitting the comma from these sentences: (i) *She hoarded silver, paper and rags.* (ii) *We ate chocolate, cakes and ices.* Hyphens can be used to avoid ambiguity if *silver-paper* and *chocolate-cakes* are intended.

4. **To separate parts of compound or complex sentences, to aid comprehension by separating different ideas:** *Although his grant had run out, he bought an expensive computer.* Do not, however, use a comma to separate the subject or the complement from the verb, unless a commenting section comes between them. Wrong: *Such absurd, extravagant and distracting gestures, should not be used when speaking to a very small audience.* The second comma is wrong as it hinders the flow of meaning from the subject, *gestures,* to the verb phrase, *should not be used.*

5. **To separate sentence modifiers such as *moreover, indeed, however*:** (i) *The wounded predator, however, continued its attack.* (ii) *Indeed, we have never had a clearer set of gels.*

6. **To separate parts of dates and addresses, and in opening and closing letters:**

28, The Terrace,
London, SW19 6PY
14th October, 1993

Dear Peter,
Thank you for your invitation. The answer is 'Yes, please!'

Yours sincerely,
Jacob

Some or all of those commas in the address and date are now often omitted.

7. **To separate the figures within a number into groups of three, from the right-hand end if there is no decimal point, or from the decimal point, going to the left only:** *6,457; 13,109,896; 4,678.98577; 0.9876547*. This helps the reader. In some foreign countries, such as Italy and Spain, a comma is used instead of a decimal point, *e.g. 9,3* for *9.3*.

8. **To separate two independent clauses (which usually could be written as separate sentences) which are joined by a co-ordinating conjunction** (*e.g. and, but, or, nor, so, yet, either. . .or, neither. . .nor*): *It is necessary to eat, but it is better to combine necessity with pleasure.*

9. **To prevent misreading, even temporary misreading:** *If you want to shoot the farmer will lend you his gun.* This would be clearer as: *If you want to shoot, the farmer will lend you his gun.* The first version initially implies shooting the farmer.

10. **To show the omission of a word or words whose meaning is understood:** *He can identify plants; she, insects.*

One of the most common errors is to use only a comma, without a conjunction, to join main clauses which could stand as separate sentences, both having a subject and a finite main verb (not just an infinitive such as *to be* or a participle or gerund such as *being*). Wrong: *We identified both the enzymes, they worked best at low pH.* Two such 'sentences', if linked, should be joined by a stronger link than just a comma. A semicolon, or a comma plus a linking conjunction, should be used. Use a colon if the second 'sentence' explains, expands or summarises the first one.

The Semicolon [;]

The semicolon is an important but often under-used punctuation mark. It is particularly useful in long, complex sentences, giving a longer pause than a comma, but not as long a pause as a full stop. There are several major uses:

1. **Semicolons can be used instead of commas to separate items in a list,** especially where some items are long or contain commas themselves, or to avoid misunderstandings: *At the zoo we saw a brown bear, which was suckling two tiny cubs; a sleepy crocodile; two stick insects, each looking like a dead twig; and five elephants.* Here, having a semicolon, not a comma, after *cubs* avoids any implication that the bear was suckling all remaining items in the list. A comma is usually sufficient before the last item, but a semicolon here makes it clear that the stick insects did not look like five elephants as well as like dead twigs.

2. **To separate clauses which could have been two different sentences, but which are closely related in meaning, and are of similar importance:** *It was long past midnight, in an otherwise deserted building; she shivered with fear.* Although one could use a full stop after *building,* making two sentences, joining the sentences with a semicolon shows better that her action was related to, not independent of,

the time and place.

Two statements joined by a semicolon may provide contrasting ideas: *The very young often wish to be older; the very old would prefer to be younger.*

The second or later statement may complement the first one (a colon could be used instead of a semicolon here): *The isolation of the mitochondria seemed unusually easy; the new centrifuge had been costly but effective.*

3. **To come before linking words** such as *therefore, nevertheless, however, besides,* when they join two independent clauses: *She hated Cambridge; nevertheless, she flourished there.*

A semicolon is often equivalent to, and replaceable by, a comma plus an appropriate conjunction: (i) *She liked Robert; he disliked her.* (ii) *She liked Robert, but he disliked her.* With this equivalence, some textbooks state that it is wrong to have a semicolon followed by an ordinary conjunction (*e.g. and, but, for, nor*), but other books permit it. The first semicolon example above, about the zoo, shows a case where a semicolon followed by *and* is useful.

The Colon [:]

A colon is generally a punctuation mark of introduction, signalling 'look ahead', rather than of separating or stopping things. It is used:

1. **To introduce a list:** *Check that you have everything you need: microscope, slides, stains and mounting materials.*
2. **To introduce direct speech:** *He said: 'These bacteria are dangerous.'*
3. **To introduce an explanation, expansion or summary of the first part of a sentence:** *There were two problems: his small grant and his need for expensive isotopes.*

There are some occasions when either a colon or a semicolon could be used to join two sentences, but choose a colon if the second one expands, explains or summarises the first one, with the colon signalling 'look ahead!': *At last he told us the source of his funding: the army had paid for his research on human neurotoxins.*

The Brackets [()]

Brackets are always used in pairs, to separate supplementary, subsidiary or explanatory material from the main flow of a sentence: *Visitors arriving for the conference in Glasgow on 2nd January (a Bank Holiday in Scotland) should make their own arrangements for lunch that day.*

The material inside the brackets can be referred to as being 'in parenthesis'. In equations, there may be different types of brackets, ( ), [ ], { }, to show different hierarchies of terms. Brackets are also used to enclose references, interruptions and afterthoughts: *Mr Brown's comments (letter, The Times, Aug. 3) show a total ignorance of female psychology.*

Brackets make a firmer separation of the enclosed material than do two commas. If the words in brackets come at the end of a sentence, a full stop (or [?] or [!]) comes after the second bracket: *Punctuate your writing carefully (it will help you and your readers).* If the words inside the brackets are a separate, complete sentence, put a full stop (or [?] or [!]) before the second bracket: *We had the paper published. (But the editor asked for many changes.)*

Square brackets, [ ], are used to enclose editorial comments or explanations in material written by a different author: *Rachael [his second wife, who died in 1988] had given James the initial idea for the project.*

The Dash [–]

A single dash is used:

1. **To mark a pause for effect:** *He wrote three versions of the manuscript – but all were rejected.*

2. **To introduce an afterthought, a summary, an elaboration or a change in direction of thought:** *'I was working on animal camouflage during the war – but I mustn't bore you with ancient history.'*

A pair of dashes is used to show an interruption in the flow of thought, to enclose a side comment or a subsidiary idea: *His supervisor – a brilliant biochemist in his earlier years – encouraged him to apply for the job at Beecham's.*

Pairs of dashes, brackets and commas are sometimes interchangeable, but may give different emphasis.

The Hyphen [-]

While the dash indicates separation and has a space on either side of it, the hyphen has no surrounding spaces and is a joining mark within words and compound expressions: *ex-wife, short-sighted, blue-eyed, do-it-yourself.* A hyphen is particularly useful to distinguish words with the same letters but different meanings: *To re-cover the chair with velvet; to recover the chair from the rubbish dump. To resign; to re-sign.* Hyphens are very valuable for avoiding ambiguity: *An ancient-history teacher; an ancient history-teacher; enzymes which re-fuse the matching ends* (*refuse* has a different meaning); *extra marital sex, extra-marital sex; extra nuclear DNA* [additional nuclear DNA], *extra-nuclear DNA* [DNA outside the nucleus]. A hyphen is often used if combining two words, or a prefix or suffix with a main word, would result in two identical vowels or three identical consonants coming together: *pre-emptive; co-opt; grass-seed.*

Hyphens are also used when writing numbers, and fractions used as adjectives, if consisting of more than one word: *By the age of twenty-three, he had spent his one-third share of his father's legacy.* Note that *one third of the class* would not have a hyphen after *one,* as *one third* is then a noun phrase, not an adjectival phrase. *The man with many children paid thousands of pounds for their twenty first birthday parties* is ambiguous: it needs a hyphen after *twenty* or after *first,* depending whether '*21st birthday*' or '*20 first-birthday*' is intended.

The hyphen is used to show that a word has been divided at the end of a line. Do not divide one-syllable words and do not divide a word so as to leave only one letter before or after the division. In general, divide words at the ends of syllables (pronounce them, if unsure where the syllables end, *e.g. mi-cro-scop-ic*). Avoid distracting fragments, as in: *the-rapist* or *depart-mental.* If using a word-processor, be careful about using hyphens at the end of lines, as later editing could move the hyphenated word away from the end of the line, but retain the hyphen. Consult the manual for advice on 'hard' and 'soft' hyphens.

The Apostrophe [']

This has several uses:

1. **To indicate that a letter or letters have been omitted:** *don't* (do not); *I'll* (I will); *it's* (it is or it has – note that the possessive pronoun *its* does **not** have an apostrophe).

2. **To form plurals of expressions with no natural plural:** *The 1980's were a better decade for our research.*

3. **To form plurals of letters** (*There are two c's and two r's in 'occurred'*), **numbers** (*Two 33's make 66*), **symbols** (*The equation contained two x's*) **and of words referred to as words** (*Two but's in one sentence can sound awkward*).

4. **To form the possessive case of a noun:**
   (i) With a singular noun, add an apostrophe and an *s* to the basic form: *John, John's essay; the tiger, the tiger's teeth.* One can omit the *s* if it makes an awkward combination of *s*-sounds: *James's house,* but *James' serious suspicions.*
   (ii) With a plural noun, add only the apostrophe if the plural ends in *s* already: *the two microbiologists' successes; the books' prices; the ladies' hormones.*
   (iii) If the plural noun does not end in *s* already, add an apostrophe and an *s*: *the men's choice; the people's reactions.*

Do not use an apostrophe in possessive pronouns: *its, hers, his, ours, yours, theirs.*

Do not use an apostrophe in the plurals of ordinary words which are not possessive. Wrong: *Cheap cauliflower's! Bargain shirt's.* These are ordinary plurals, with no sense of possession, whereas *The shirt's price* does need an apostrophe, showing possession of the price by the shirt.

The Inverted Commas [" " or ' ']

Inverted commas are also called quotation marks, speech marks or quotes. The closing quotation mark comes after any punctuation mark which is part of the quoted section, but before any mark which is not: *They asked: 'Is it a bear skull?'*. Where a sentence being quoted is broken by the subject and verb of saying, the first part of the quotation has a comma before the quotation mark: *'I shot the golden eagle,' he said, 'because it was killing my lambs.'*

1. **They are used in pairs to enclose direct speech;** that is, the exact words spoken: *'We've won!' she shouted.* When the words spoken come before the verb relating to the act of saying, they are followed by a comma (or [?] or [!]) before the inverted comma, even though the spoken sentence has ended: *'I'm going to France soon,' she declared.* The first word spoken has a capital letter. Where the spoken sentence is broken by the subject and verb of saying, the continued speech after the break is still the same spoken sentence and so continues with a small letter (unless the first word is a proper noun): *'It is my wish,' the lecturer said, 'to entertain as well as to instruct you.'* In quoted speech lasting more than one paragraph, there is an initial quotation mark and one at the beginning of each subsequent paragraph (to indicate that the speech continues), but no quotation mark at the end of each paragraph (the speech has not finished), until the final closing quotation mark.

   Although sets of either single or double inverted commas can be used – different publishers have different conventions – using double inverted commas avoids possible confusion when an apostrophe follows a final *s:* (i) *He said: "It is Anne's".* (ii) *He said: 'It is Anne's'.*

2. **Inverted commas are used in pairs to indicate direct quotations in writing,** which is their main use in written biology. For quotations within quotations, one alternates single and double inverted commas:

The professor wrote that *'My happiest times as an ecologist were with "the boys in the bush" in North Africa'.*

3. **Inverted commas can also be used to enclose slang, dialect or foreign expressions, and technical terms or other terms that seem inappropriate to their context:** (i) *He complained to the doctor of a 'runny tummy',* [slang for diarrhoea]. (ii) *Her natural 'joie de vivre'* [French for 'joy of living'] *was accentuated by several glasses of champagne.*

4. **Inverted commas are sometimes used to enclose the titles of books, plays, poems, newspapers, etc.,** when quoted in writing, but such marks are often omitted now. In some cases, italics are used instead of quotation marks for titles, slang and foreign expressions.

The Slash [/]

The slash is also known as the solidus, the slant, the oblique and the stroke. It is used in fractions and divisions: *3/5ths of the distance*; in dates: *21/12/93*; to show alternatives: (i) *Your coach/train/boat/plane ticket;* (ii) *He/she should give blood only if. . . .*

The Capital Letter

A mistake over capital letters can be misleading. The Minutes of a staff meeting included: *The division of Biology, which we discussed previously. . . .* That implies splitting up Biology, whereas a larger grouping, a Division of Biology, was intended. Take care where words have different meanings with and without a capital letter, especially if the word comes at the beginning of a sentence: *Swedes are a favourite food in Norway* might be considered ambiguous.

Capital letters, sometimes called upper case letters, are visual symbols, helping the reader in the same way as punctuation marks. They are used in many places, including: the beginning of a sentence; in proper names (*Liverpool, Joan Smith, River Taff, North Pole, Fleet Street*); in titles, for example, of people, plays, books and journals (*the Vice Chancellor, the Managing Director, A Textbook of Biochemistry, Genetical*

Research); in days of the week; in months and in the pronoun *I*.

Capitals are not used for the seasons (*spring, winter*), nor for points of the compass unless part of a proper name: *Go east, to East Sheen.* In titles, less important words such as *a, of,* or *the,* often do not have a capital letter. Be careful not to use a capital letter for a word which has a capital when part of a title, but is not being used as part of a title: *This morning Professor Roberts opened the new Genetic Fingerprinting Laboratory. The **p**rofessor hoped that the work of the **l**aboratory's staff would help in the fight against crime.* In the second sentence, *professor* and *laboratory* are not parts of titles.

In scientific Latin names, by international convention, the genus always has a capital letter and the species always has a small (lower case) letter, even if based on a proper name: *Salmonella typhimurium; Saccharomyces carlsbergensis*; *Rosa chinensis* (small initial letter for the species, in spite of the name meaning *from China*); *Homo sapiens.* Newspapers and students frequently get such capitals and lower case letters wrong, although the rules are so simple. Scientific names are normally printed in italics, or can be written with underlining.

In English, but not in some other languages, adjectives of nationality normally have a capital letter: *a French iris; an American university; German measles.* Except at the beginning of sentences, capital letters should not be used for chemicals, elements or enzymes, unless they are trade names or proper names: *sodium chloride, oxygen, urease, Terylene®, Difco Bacto-agar.* In symbols for chemical elements and dominant alleles, only the first letter is a capital, *e.g.* NaCl, Wx.

Italics and Bold Print

Word processors make it easy for students to use these print devices for special emphasis, as we frequently do in this book. They can be used to make headings or subheadings stand out, or to distinguish different types of text within a sentence, such as distinguishing examples from the text commenting on those examples, or even to draw attention to single letters, as in the above section on capital letters: *The **p**rofessor hoped that the work of the **l**aboratory's staff would help in the fight against crime.* In this book, we show three methods of making quoted examples stand out from text commenting on them, but normally one should be consistent within a piece. Here we have used quotation marks, italics and a different font (typeface) at different times.

Make sure that your printer can deal with these kinds of print. A colleague used bold print to avoid ambiguity, referring to two MSc options: **Applied Entomology** and **Plant Pathology and Nematology**. This was eventually printed without bold type, so on the proofs I (BL) inserted a comma to make the correct combination clear: *Applied Entomology, and Plant Pathology and Nematology.*

Punctuation Exercise

Now spot the punctuation (and spelling) errors in the following actual examples of students' work. Answers are given on page 252.

1. Bacterial cell walls have great medical significance, they are responsible for bacterial virulence.
2. Only mycoplasmas, the smallest bacteria lack a cell wall.
3. If non nutrient requiring genes are transduced into nutrient requiring bacteria co transduction can be found by nutrient requiring tests.
4. Penicillin inhibits the enzyme transpeptidase; which is involved in the formation of cross-links between glycogen strands.
5. Transformation requires no cell to cell contact the DNA which will be intergrated is in the media.
6. The child has problems coping with the hormone's in the mothers blood.
7. Due to differences in codon bias difficulties arise at translation, with a higher frequency of misstranslation.
8. It seems likely that without aphids parthenogenetic reproduction insecticide resistance would spread less rapidly.
9. If this occur's three or four embryo's should be transferred. . . .

CHOOSING THE RIGHT WORD

Many words look similar, or sound alike (homophones), and are easily mistaken for each other. Other words look quite different, but people confuse their meanings. If you choose the wrong word, it gives the appearance of ignorance and can mislead the reader or cause scientific nonsense. Always check with a dictionary if you need help, and devise your own ways of remembering which word has which meaning. Beware of using a term from one area of biology in another if confusion could result; for example, *the dominant type of plant* in ecology would mean the most frequent type, but in genetics it would mean the type with the phenotype showing the dominant, not the recessive, allele.

- ***affect/effect.*** When a student wrote: *A bad diet effects a woman's pregnancy,* he wrote scientific nonsense, which deserved to lose marks for being bad science. *To effect* is to accomplish fully, to bring about, while *to affect* is just to have some influence on. A bad diet can indeed *affect* (have some influence on) a woman's pregnancy, but it cannot bring about (*effect*) a pregnancy. About 46 per cent of our (BL and JP) students confuse affect/effect, so make sure that you know the difference.

- ***assume/deduce.*** You may make an assumption or supposition on theoretical grounds, but if you conclude something from actual data, then you are deducing, not assuming.

- ***biannual/biennial.*** *Biannual* usually means twice a year, while *biennial* means happening every two years, or lasting two years, or taking two years, such as a plant which only flowers in its second year, then dies.

- ***complementary/complimentary.*** Many students write of *complimentary genes* instead of *complementary genes.* It is easy to remember the meanings as *complementary* relates to *completing* each other's action, while to *compliment* is to say pleasant things about someone.

- ***fewer/less.*** *Few* **and** *fewer* refer to numbers (discontinuous variables), while less refers to amounts (continuous variables): *he used fewer rats* (**not** *less rats*); *smaller ants carry less weight of leaf to the nest.* Think of the different implications of: *fewer beautiful women* **and** *less beautiful women.*

- ***infer/imply.*** Only an intelligent being can *infer*, make an inference or deduction from data; data or experiments can never infer anything, although they can *imply*, meaning to give rise to an inference, to suggest indirectly.

- ***principal/principle.*** *Principal* means 'chief' or 'head', and can be an adjective or noun; the noun *principle* is a rule, a fundamental truth. The difference is easily remembered from *My pal, the Principal, has multiple principles.*

- ***stationary/stationery.*** It is easy to remember that *stationery*, with *e* near the end, includes *envelopes*, with *e* near the end, while 'the *car* was *stationary*' could help you to remember which word means 'not moving'.

- ***their/there.*** It is amazing how often intelligent undergraduates confuse these elementary words: *their* is the possessive of *they*, containing the letters of *heir*, an intended possessor; *there* is the opposite of *here*, the letters of which it contains.

- ***weather/wether/whether.*** An educated biologist should never confuse *weather*, to do with climatic conditions, with *wether*, a castrated ram, or *whether*, introducing alternatives. Should you need a mnemonic (memory device), *weather* includes the mixed-up letters of *heat*, and the other two words do not.

- ***were/where.*** *Were* is the past tense of *are*; *where* refers to which place, and contains the letters of *here.*

It astonishes me (BL) that students make such confusions as: *Only the top 4 per cent of* ***bores*** *are allowed to breed [**boars**] ; pure-**bread** rabbits [**bred**]; a **holy** tree [**holly**]; **sweat** peas [**sweet**]; **pair** trees [**pear**]; **larger** [**lager**] beer.* Some of these errors were consistent, as in *locus/locust,* *callous/callose, horse/hoarse,* not just 'a slip of the pen'. It is alarming that many student nurses now put 'to *elevate* the disease', to increase it, when they mean 'to *alleviate*', to make it less bad. One PhD thesis consistently had *proscribe* [forbid] for *prescribe* [to require, to lay down rules or make out a prescription] and *leech* for *leach*, changing the meanings drastically.

Look at the following list of words which students have confused, and use a dictionary if necessary to make sure that you know **all** the distinctions: *accept/except; affected/infected; base/bass; boar/bore/boor; compliment/complement; confirmation/conformation; denote/donate; discrete/discreet; fortuitous/fortunate; its/it's; media/mediums; moral/morale; offspring/of spring; on/one/won; preyed/prayed; principal/principle; right/rite/write; simulate/stimulate/assimilate; raped/rapped; site/sight; stripped/striped; thymine/ thiamine; to/two/too.*

Do build up your general and technical vocabulary, by using a dictionary to look up any unfamiliar words you read or hear. A survey (J. Richardson and R. Lock, *Journal of Biological Education*, 1993, 27: 205-212) found that a substantial proportion of British A-level Biology candidates had trouble understanding the meanings of some words used in A-level exam papers. If you do not understand the words in your university exam papers, you will have trouble with the answers too.

SPELLING

Like word confusions, bad spelling gives an impression of carelessness and ignorance, and can lead to confusion and misunderstanding. Unless you have a medical condition such as dyslexia, you can be a good speller, and it is never too late to learn. My (BL) spelling was poor until my late twenties, when frequent and justified criticism from a Sri Lankan research student made me take action. I started learning rules, using a dictionary much more, and playing word games such as crosswords, where wrong spellings were detectable. Learning a few rules will enable you to avoid many common blunders such as *occured, assymetry, unaturally, definate, acheive, innoculate, normaly, begining,* and *dissapear.* Do not worry if some of the rules look a bit complicated at first, with mentions of vowels, consonants, syllables and stresses (all defined above), because the examples should make the rules clear. Even if you do not remember the rules, you can often use the pattern of a known word to find the rule to apply to a word you have doubts about. Although there are some exceptions to some rules, especially as *l, w* and *x* sometimes follow different rules from those for other consonants, the rules are still worth learning.

Some Useful Rules

1. **Pronounce the word carefully, and use special 'mental pronunciations' to help remember words you find difficult.** In your mind, you could pronounce and stress the second *n* of *environment,* the *p* of ***psy**chology,* the *k* of ***k**nob,* or the *g* of ***g**nat.* Pay attention to long and short vowels in your pronunciation, as they may help you. A short vowel often comes before a doubled consonant, especially if the consonant is not *l,* so I use this when spelling *accommodation* and *accumulation,* as the initial short vowel *ă* is followed by a double *c,* then there is a double *m* after the short vowel *ŏ* in *ăccŏmmōdātion,* and a single *m* after the long vowel *ū* in *ăccūmūlātion.* A long vowel sound followed by a consonant (not *w, x* or *y*) often indicates a silent final *e* after the consonant: *rŏb, rōbe; dĭn, dīne.*

2. **Keep a record of words you find difficult to spell,** perhaps in an address book under the relevant initial letter. Write them out to help learn the spelling, and devise a mnemonic if possible.

3. **Use your knowledge of related words if troublesome unstressed vowels are not pronounced, or are pronounced unclearly.** Different unstressed vowels may be pronounced very similarly, causing doubts about their spelling, as in the last vowel of *emigrant.* The *a* is a vague short sound which could be *ĕr, ă, ĕ, ŭ,* or *ŭr.* The very clear stressed *ā* sound in the related word *emigr**ā**tion* helps you to guess the unstressed *ă* in *emigr**ă**nt.* Similarly, the unstressed vowel before *-nce* in *substance* and *essence* can be worked out from stressed, clearly pronounced vowels in the related words *substantial* and *essential.* Use this method now to find the missing vowel [-] in: *defin-te, radi-nt,*

irrit-nt, environm-nt, degr-dation. See answers on page 252.

4. **Is it *ie* or *ei*?** The rule is: '*i* before *e* except after *c*, if it rhymes with *bee*.' This helps with: *receive, achieve, believe, chief, receipt, deceive, brief,* etc. The few exceptions include *protein, caffeine* and *seize.* Most, but not all, *ie* or *ei* words where the vowel sound does not rhyme with *bee* have *ei: deign, eider, feint, heifer, heir, reign, surfeit, vein, weigh,* but note *friend.*

5. **Is it *c* or *s*?** By remembering *the advice* and *to advise,* where the noun *advice* and the verb *advise* are pronounced differently, you can work out other cases where the noun has *c* (*the practice, the licence)* and the verb has *s* (*to practise, to license),* even when the pronunciations do not differ.

6. **Prefixes are usually added to the base word without changes to either:** *dis+solve = dissolve; dis+appear = disappear; mis+spell = misspell; inter+related = interrelated; inter+act = interact; un+natural = unnatural; in+oculate = inoculate.* The few exceptions usually involve *l*: *all+together = altogether; well+come = welcome.*

7. **Suffixes.**
   (a) Suffixes beginning with a consonant ('consonant suffixes'), as in *-ly,* do not usually change the base word: *normal+ly = normally; complete+ly = completely; govern+ment = government; hope+ful = hopeful.*
   (b) Suffixes beginning with a vowel ('vowel suffixes'); *y* counts as a vowel here:

   (i) If the base word ends in a silent *e,* you generally drop the final *e: hope, hoping; simple, simply; complete, completion.* The silent *e* is retained, however, if needed with a final *g* or *c* to keep the pronunciation of that consonant soft before suffixes beginning with *a* or *o,* as those vowels – unlike *e* and *i* – do not soften those consonants: *notice, noticeable* (*noticable* would be pronounced *notickable*); *outrage, outrageous.* The final *e* is also retained to prevent confusions: *dye, dyeing; die, dying.* Exceptions are: *truly, duly, argument.* Try these rules on *clone+able; peace+able; forgive+ing; manage+able; manage+ing;*

stone+y; code+ing. See answers on page 252.

(ii) In words with one syllable, and a single final consonant after a single vowel, you double the consonant when adding the vowel suffix: *hop, hopped; slim, slimmer; plan, planning* (without knowing such rules, one might put *hoped, planing,* changing the meanings).

(iii) In words of one syllable ending with two consonants, or with two vowels together before a final consonant, you do not double the final consonant: *harp, harping; cool, cooling.*

(iv) In words of two or more syllables ending in a single consonant preceded by a short vowel, you do not double the final consonant if the stress is on the first syllable: *al'-ter, al'-tered; of'-fer, of'-fering;* if the stress is on the final syllable, you double the final consonant: *be-gin', be-gin'-ning; sub-mit', sub-mit'-ted; re-fer', re-ferred'.* Interestingly, if adding the suffix changes the stress pattern, go by the stress pattern in the final word rather than that in the base word: *re-fer',* but *ref'-er-ence,* with a single *r* before the suffix because of the shift in stress. Note that in words ending in *l* preceded by a short vowel, whether stressed or not, the *l* is usually doubled in Britain (but not in the USA): *tra'-vel, travelled; com-pel', compelled.* Try this very useful rule (iv) now, adding *-ed* to: *disbud, occur, differ, transmit, target, order, prefer, propel, commit, label.* Pronounce the words if you are unsure which syllables are stressed, or try stressing each syllable in turn to find which version sounds right. See answers on page 252.

(v) Where the word ends in a single consonant preceded by two vowels, do not double the consonant: *remain, remaining; unveil, unveiling.*

(c) If a word ends in a consonant and *y*, change the *y* to an *i*, whether using a consonant suffix or a vowel suffix (unless the vowel is *i*, when the *y* is unchanged, to avoid a double *i*): *hazy, hazily; happy, happiness; marry, marriage; carry, carrying.*

(d) If a word ends in *y* preceded by a vowel, the *y* is unchanged: *employ, employed, employment.*

Try these for yourself: *necessary+ly; lazy+ness; display+ed; friendly+est; bury+ing.* See answers on page 252. Easy, isn't it? From these rules on prefixes and suffixes, you can actually work backwards to find the appropriate base word, if necessary. If you were not sure which of *robing* or *robbing* came from *to robe*, you could use the above rules to work out *robe+ing* and *rob+ing*, which would give *robing* and *robbing*, respectively, solving your problem and avoiding an error.

8. **Breaking words down into their component parts can often help with their spelling**, including using your knowledge of prefixes and suffixes: *government = govern+ment; misspell = mis+spell; statistician = statistic+ian; conscience = con+science.*

Learning Word Origins

You do not need to know Latin (L) or Greek (Gr) to learn the origin of words used in biology, but learning a few of the common roots, whether prefixes, suffixes, or base words, is extremely useful. It helps with the spelling (*anti-* or *ante-natal*? *forhead* or *forehead*? *phylum* or *phyllum*?) and with the meaning. For example, if you look up the origin of *pachyderm*, a thick-skinned animal such as an elephant, you find that it comes from *pachys* (Gr) meaning thick, and *derma* (Gr) meaning skin. That knowledge helps when you come to *pachytene* in meiosis, when the chromosomes are thick threads (Gr *tainia*, a band or thread), or *hypodermic*, under the skin (*hypo*, Gr, under). Use a good dictionary to find word origins, and you will soon find that you can guess the meanings of many new words just from knowing their components. You do not need to know which languages the roots come from, although when coining new words, purists like to combine say two Greek roots or two Latin roots. The list that follows includes some of the more common ones, and you can easily make a list of common ones in your specialist subject area.

Word-root	Meaning	Examples
a-	not, without	asexual
ab-, abs-	away from, off	abaxial (away from the axis), absent
acro-	apex	acrocentric (centromere near the end of the chromosome)

Word-root	Meaning	Examples
ad-	near, to, at	adaxial (towards the axis)
aero-	air	aerobic
allo-	different, other	allopolyploid (polyploid with sets of chromosomes from different species)
amphi-	both	amphibious (living both in water and on land)
an-	not	anaerobic
ana-	up	anabolism (building up metabolites)
andr-/andro-	male	androgen (male hormone)
annul-	ring	annulus (ring-shaped structure)
ante-	before	antenatal (before birth)
anth-	flower	anthesis (flower opening)
anti-	opposed to	antiseptic
arch-	first, chief, primitive, original	archaeology, Archaea
aur-	ear	auriculate (ear-shaped)
aut-	same, self	autogamy (self-fertilisation)
bi-	twice, double	bigamy (see gam-)
bio-	life	biology
brachy-	short	brachypterous (short-winged)
carp-	fruit	monocarpic (flowering only once)
cata-	down, away	catabolism (breaking down metabolites)
caul-	stem	acaulous (stemless)
centi-	hundred, hundredth	centipede, centimetre
chloro-	green	chloroplast
chromo-	coloured	chromatin (stainable DNA structures)
chron-	time	chronology
-cide	killer	insecticide
circum-	around	circum-polar
cis-	on the same side	Cisalpine, cis-isomer

Word-root	Meaning	Examples
clad-	branch	*Cladophora* (a branching alga)
clav-	club	clavate (club-shaped)
cleist-	closed	cleistothecia (closed fungal fruit bodies)
coccus-	round body, berry	*Streptococcus*
contra-, counter-	against, opposite	contradict, counter-current
crypto-	hidden	cryptic
cyano-	blue-green	Cyanophyta (blue-green algae)
cyto-	cell	cytology
deca-, dec-	ten	decapod (having 10 legs)
deci-	one tenth	decimal, decimate
delt-	triangular	deltoid muscle
demi-	half	demi-wolf (dog-wolf hybrid)
-derm	skin	hypodermic
di-	two, twice	diploid (two sets of chromosomes)
dia-	across, apart	dialysis, diakinesis (stage of meiosis when the centromeres start to move apart)
dys-	bad	dysfunction
ecto-	external	ectoderm
endo-	inner	endoderm, endonuclease
epi-	on, over	epidermis
equi-	equal, level	equilibrium
eu-	true, good, well	eukaryotic (with true nuclei)
ex-	out of, former	exhale, ex-conjugant
extra-	more, outside	extra-terrestrial
for-	prohibit	forbid
fore-	front, before	forehead, forecast
gam-	union, marriage	gametes
gastro-	stomach	gastric, gastropod
-gen	producing	mutagen (producing mutations)
geo-	earth	geology

Word-root	Meaning	Examples
gon-	angle	trigonous (three-angled)
gymn-	naked	Gymnosperm (seed naked, not in a carpel)
gyn-	female	gynaecology
haem-	blood	haemoglobin
hepta-	seven	heptamerous (parts in sevens)
hetero-	unlike	heterogeneous
hexa-	six	hexagonal
homo-	alike, same	homogeneous
hydro-	water	hydrophobia (fear of water)
hyper-	excessive, over	hyperactive
hypo-	under, too little	hypothermia (too cold)
in-	not	incompatible
inter-	among, with each other	interspecific (between different species)
intra-	within	intraspecific (within one species)
iso-	equal, alike	isogamy (fusion of similar-sized gametes), isobar
karyo-	kernel, nucleus	karyogamy (nuclear fusion)
kilo-	thousand	kilometre
kin-	movement	kinesis, kinetic
lacto-	milk	lactose (milk sugar)
lepid-	scale	Lepidoptera (scales on wings)
lepto-	slender	leptotene (stage of meiosis with very slender chromosomes)
lign-	wood	lignin
ligul-	strap, tongue	ligulate (strap-shaped)
litho-	stone	monolith
-logy	subject, study of	biology
macro-	large	macroscopic
mal-	bad	malformed, malignant
mega-	large, million	megawatt
mer-	part	pentamerous (parts in fives)

Word-root	**Meaning**	**Examples**
meso-	middle	mesophyll
micro-	small	microscopic
milli-	thousand, thousandth	millipede, millimetre
mono-	one	monogamous, monosaccharide
morph-	shape, form	morphogenesis
multi-	many	multiple
myc-	fungus	mycology
neo-	new	neonate (newborn)
neuro-	nerve	neurology
non-	not	non-toxic
octo-	eight	octopus
oec-, ec-	house, home	monoecious (male and female parts on one plant), ecology (study of organisms' 'homes')
oligo-	few	oligonucleotide (a few nucleotides joined)
omni-	all	omnivorous (eating all types of food)
ortho-	straight, correct	orthodox
pachy-	thick	pachyderm (thick-skinned animal)
paleo-	old	paleontology
pan-	all	panacea (cure-all), panmixis
path-	disease	psychopath, pathology
penta-	five	pentagon
peri-	around	perimeter
phil-	loving	thermophilic (heat-loving)
photo-	light	photosynthesis
phyc-	alga, seaweed	phycoxanthin (an algal pigment)
phyl-	tribe	phylum
phyll-	leaf	phyllotaxis (leaf arrangement up stem)
phyt-	plant	xerophyte (plant of dry places)

Word-root	Meaning	Examples
-ploid	number of sets, *e.g.* of chromosomes	triploid
-pod	leg	tripod
poly-	many	polysaccharide (made of many sugar units)
post-	after	post-mortem (after death)
pre-	before	prenatal (before birth)
pro-	before, for	proplastid (body which differentiates into a plastid)
pseudo-	false	pseudopodia (false feet)
psycho-	mind	psychopath
quad-	four	quadruped
quasi-	as if, like	quasi-neutral
re-	back	recall
ren-	kidney	renal
saccharo-	sugar	polysaccharide (sugar of many single units, *e.g.* starch)
schizo-	split	*Schizosaccharomyces* (fission yeast – 'splitting sugar[-loving] fungus')
semi-	half	semicircular canals
sex-	six	sextuple
som-	body	somatic tissue
sub-	under	submarine
super-, sur-	above, too much	superhuman, surfeit
syn-, sym-	together, same	synchronous, symmetric
tetra-	four	tetrahedral
thec-	case, sheath	perithecia (fungal fruit bodies encasing the sexual spores)
thermo-	heat	thermophilic
trans-	across, on the other side, beyond	transpose, transport
tri-	three	tripod
trop-	turn	tropism

Word-root	Meaning	Examples
troph-	nutrition	phototrophic (using light energy to build up foods)
-vorous	eating	carnivorous (flesh-eating)
ultra-	beyond	ultraviolet light
uni-	one	unicellular
zoo-	animal	zoology
zyg-	yoke, union	zygote (result of union of gametes), zygotene (stage of meiosis when chromosomes pair)

Words Frequently Misspelled in Biology

This is only a very small selection! Keep a list of words you often misspell, and learn from it.

- *abbreviation*
- *aberration*
- *accommodate*
- *accumulate*
- *achieve*
- *acquire*
- *aerial*
- *airborne*
- *all right* (**not** *alright*)
- *apparatus*
- *aseptic* (**not** *asceptic*)
- *asymmetry*
- *auxiliary*
- *basically*
- *beneficial*
- *biennial*
- *buoyant*
- *calendar*
- *committee*
- *conscientious*
- *deceived*
- *decrepit*
- *definitely*
- *degradation*
- *deleterious*
- *descendant*
- *desiccate*
- *developed*
- *disappear*
- *disappointing*
- *discipline*
- *Drosophila*
- *efficient*
- *eighth*
- *embarrassed*
- *emigrate,*
- *endeavour*
- *environment*
- *Escherichia*
- *exaggerated*
- *exceed*
- *extraordinary*
- *extreme*
- *fascinate*
- *fluoride* (**not** *flouride*)
- *foreign*
- *forty*
- *fridge*
- *fulfil*
- *gauge*
- *height*
- *hygiene*
- *immediately*
- *immigrate*
- *inadvertently*
- *incompatible*
- *independent*

- *indictment*
- *indispensable*
- *inoculate*
- *irrelevant*
- *its* (possessive pronoun)
- *knowledgeable*
- *liquefied*
- *literature*
- *lysine*
- *maintenance*
- *manoeuvring*
- *media*
- *miniature*
- *miscellaneous*
- *necessary*
- *noticeable*
- *occasionally*
- *occurred*
- *occurrence*
- *paralleled*
- *perseverance*
- *personnel*
- *pigeon*
- *possesses*
- *potato*
- *potatoes*
- *precede*
- *prejudice*
- *presence* (**not** *prescence*)
- *privilege*
- *procedure*
- *proceed*
- *programme* (**but** *program* **for a computer**)
- *psychiatrist*
- *pursue*
- *putrefied*
- *receive*
- *reciprocal*
- *recommend*
- *schedule*
- *sensitive*
- *separate* (**not** *seperate*)
- *succeed*
- *successful*
- *succession*
- *supersede*
- *symmetry*
- *technique*
- *technical*
- *tomato*
- *tomatoes*
- *transference*
- *transferred*
- *unnatural*
- *unnecessary*
- *until*
- *usually*
- *vegetable*
- *veterinary*
- *vigorously*
- *visible*
- *weight*
- *weird*
- *wholly*
- *yield.*

SOME USEFUL GRAMMAR

Subject/Verb Agreements, Including Collective Nouns

1. Singular subjects take singular verbs and plural subjects take plural verbs, obviously, but there are various areas of uncertainty. Newspapers and students make frequent mistakes in this aspect of English, usually because they have not bothered to identify the actual subject of the verb, especially when the verb is separated from its subject by a descriptive phrase. The descriptive phrase can always be eliminated

without losing the main sense of the sentence, but eliminating the subject spoils the sentence. For example: *One of our newest and most expensive centrifuges has broken down.* The verb is *has broken* (singular), to agree with its subject, *One*: the descriptive phrase *of our newest and most expensive centrifuges* tells us more about *One. One has broken down* still makes sense, but eliminating *One* spoils the sentence. Some people would wrongly make the verb plural, agreeing with the preceding noun, *centrifuges*, instead of the true subject, *One.*

The subject does not always come in front of the verb, so look at the whole sentence to identify the subject of the verb: *In your own hands lies the key to success. The key*, not *hands*, is the subject of the singular verb *lies*, with *In your own hands* as an adverbial phrase, answering the question *where?*

2. If you have a compound subject made up of two singular nouns joined by *and*, then you need a plural verb and the pronoun would be *they: Engineering and science, our greatest contributions to the modern world,* ***are*** *often misunderstood.* ***They*** *are. . . .* Newspapers get this wrong very frequently. The exception is when the two singular words joined by *and* are considered as one concept: *Whisky and soda* ***is*** *a popular drink;* ***it*** *is. . . .* But note the plural verb when they are separate concepts: *Whisky and soda* ***were*** *bought in separate shops;* ***they*** *were. . . .*

3. Either. . .or, neither. . .nor. If both subjects are singular, the verb is singular: *Neither chemistry nor physics* ***tells*** *us as much about ourselves as does biology.* If one subject is plural and one is singular, it sounds best if the verb agrees with the nearest preceding subject: *Either the nucleus or the chromosomes* ***are*** *visible in each cell. Either the chromosomes or the nucleus* ***is*** *visible in each cell.*

4. Each, every, everyone, anyone, either, neither, none (= not one). These are all normally singular: *Each* ***is*** *entitled to. . .; Every lecturer* ***is*** *. . .; Anyone* ***is*** *free to. . .; Neither* ***was*** *adequate. . .; None* ***takes*** *my fancy.*

5. Fewer, few. These take plural verbs: *Fewer than six* ***are*** *needed.*

6. Measurements, sums of money, percentages, etc. are considered a single entity, with singular verbs: *Two metres* ***is*** *sufficient; Five per cent* ***is*** *not enough; Three pounds* ***is*** *too dear.*

7. Relative pronouns such as *who, which* and *that* can cause trouble when the subject of verbs, as they can each be singular or plural: *Lipase is one of the enzymes which break down fats. Which* is plural here, relating to *enzymes,* not to *Lipase.*

8. You must know which words are singular and which are plural to get the verb right. See Singulars and Plurals of Nouns, pages 173–4. I (BL) hate to see: *A bacteria is. . .; The fungi is. . .; The algae has been. . . .*

9. Collective nouns – singular or plural? Collective nouns refer to groups of people or things: *class, government, committee, herd, population, couple, number, majority, the Senate, staff.* In American English they are usually treated as singular. In British English we normally use singular verbs when the collective unit is considered as or acts as a single unit, but we use plural verbs when the collective unit is considered as or acts as separate individuals. *The committee* ***has*** *reached a unanimous decision; it* ***has*** *voted to increase salaries by 10 per cent.* Here the committee acted singly. *The committee* ***were*** *arguing fiercely among* ***themselves;*** ***they*** *adjourned to* ***their*** *own offices without any agreement.* Here the committee acted individually. *A number of students* ***are*** *coming to see me over the next few days* (individual action). *The number of mistakes* ***is*** *very high.*

   Words for a class of persons usually take plural verbs: *The police* ***are. . . .***

Spot for yourselves the errors in these students' sentences. It helps to identify the subjects of the verbs. The errors are so obvious that no answers will be given, but make sure you do not make such mistakes yourself.

> The main constituent of bacterial cell walls are the lipopolysaccharides.
>
> The function of these acids are as follows. . . .
>
> A ova has. . . .

The sex of individuals are determined by sex chromosomes.

Changes in temperature has an effect. . . .

The procedure for these practicals are explained in the enclosed sheet.

This phenomena exhibits. . . . [Two errors here.]

None of the mutations were deleterious.

Inbreeding and outbreeding occurs in nature.

Singulars and Plurals of Nouns

Most English nouns have *-s* or *-es* added to the singular form to make the plural, but other nouns, including many biological words, form plurals differently and often cause students and staff trouble. Some words ending in *-f* or *-fe* have an additional *-s: proofs, beliefs, safes*; others change to *-ves: halves, wolves,* while others take either plural: *hoofs, hooves.* Words ending in a consonant plus *-y*, such as *body*, change *-y* to *-ies* in the plural: *bodies*; words ending in a vowel plus *-y* usually just have an additional *-s: monkeys, donkeys.* Words ending in *-o* usually have an additional *-s: embryos, zoos, photos, hippos, albinos,* while a few take *-es: potatoes, tomatoes.*

Some plural nouns have no singular form: *clothes, odds, proceedings,* and other words have the same form in the singular and the plural: *sheep, deer, salmon, species.* Some plural words have a different singular equivalent: *cattle, cow; people, person.*

With foreign words, the original language plural may or may not be used in English, or sometimes the foreign and English plurals are both used, as with *formulae, formulas.* Consult a dictionary if you need to. The following are singular and plural forms, respectively, of biological words: *alga, algae; hypha, hyphae; larva, larvae; vertebra, vertebrae; amoeba, amoebae* or *amoebas*; *stoma, stomata; criterion, criteria* (never write *a criteria*); *analysis, analyses; axis, axes; apex, apices* or *apexes; appendix, appendixes* (medical) or *appendices* (for books)*; skeleton, skeletons,* but *phenomenon, phenomena; ovum, ova; mycelium, mycelia; bacterium, bacteria; bacillus, bacilli; virus, viruses; fungus, fungi; genus, genera* or *genuses.*

Although you will find *data* sometimes used as singular, both authors of this book regard that as a mistake: the word *data* is the plural of *datum*. Different plurals may have different meanings: *media*, in biology, means growth media, while *mediums* means spiritualists who give seances. Do not get these confused, as many students do! Never add *-s* to existing plurals. Wrong: *stratas, bacterias, phenomenas, datas, medias.*

The Correct Prepositions

Be sure to use the correct preposition for the intended meaning. A student wrote: *a small colony **of** yellow conidia,* when the preposition should have been ***with***. Consider the differences in meaning of: *The electron was donated to/by the oxygen atom; the hyphal fragments were trapped **on/in** the filter; phosphate buffer increased the susceptibility of/to RNA viruses; inoculating the spores **in/on** the medium; entry is prohibited **by/to** the Management.* It is best to use *differs from*, not *to* or *than*, because differing has the sense of 'away from' in type, as opposed to similarities towards the same type.

Do not use unnecessary prepositions: *Where the river meets the sea* is vastly better than *Where the river meets **up with** the sea*, which is doubly redundant.

Many verbs may be used transitively and intransitively, and may need a preposition when used intransitively. For example, *I applied* [transitive] *the formula; I applied* [intransitive] ***to** the committee.* While American English permits: *I will write him,* British English needs the preposition, *I will write **to** him.*

Avoiding Incomplete Sentences and 'Hanging Phrases'

Complete sentences normally need a finite verb (see definition above), with a subject; just an infinitive or a participle is not enough: *This causing overall evolution of the population by selection.* That lacks a finite verb, having only the present participle, *causing*. It could be changed to: *This caused overall evolution. . . ,* which has a finite verb, or *this causing overall evolution of the population by selection* could probably have been joined to the previous sentence by a comma. *To go to the Research Councils for funds* is also incomplete as a sentence, having only an infinitive.

Can you see a different error here? *While looking down the microscope, the Drosophila males were seen to have sex combs on their front legs.* The part before the comma is a 'hanging phrase' – words such as *looking* are sometimes called 'dangling participles', when they 'dangle' without any referring word in the rest of the sentence. The whole phrase 'hangs' without any reference in the main part of the sentence. It was presumably not the Drosophila males who were looking down the microscope as this seems to imply. It would be better to put *When I looked down the microscope, . . .* or better still, to recast and improve the whole sentence: *When observed with a stereoscopic microscope at ×20 magnification, Drosophila males were seen to have black sex combs on their front legs (Figure 3).*

DIFFERENCES BETWEEN BRITISH AND AMERICAN SPELLINGS AND USAGES

You will notice differences in English usage between British and American textbooks and scientific journals. This is often unimportant to the reader, but if you submit a manuscript to a journal, you need to use the kind of English specified by that journal. Its instructions to authors may recommend following spellings from a particular dictionary, and editors and reviewers can be greatly irritated by an author's failure to follow instructions. Canadian journals often specify British spellings.

American English tends to omit letters which are needed in British English, as in these examples, where the British form precedes the American form. *Pipette, pipet; colour, color; mould, mold; behaviour, behavior; foetus, fetus; haemoglobin, hemaglobin; labelled, labeled; analogue, analog.*

For other words, the letters may differ or may be in a different order. Again, the British form precedes the American form: *defence, defense; tyre, tire; centre, center; litre, liter; metre, meter; titre, titer; sulphur, sulfur; aluminium, aluminum.* In British English, verb forms ending in the sound 'eyes' may be spelled *-ise* or *-ize, e.g. realise, realize; organise, organize,* while American English normally uses *-ize.* One should be consistent within a piece of writing. A few words have *-ize* only, such as *capsize,* but others only have *-ise*, such as *comprise, revise, surprise.* Check with a good dictionary.

Confusingly, some words have different meanings in British and American English, such as *corn*, which means *wheat* in the UK but *maize* in the USA; *vest* and *pants* are traditionally underwear in the UK, but mean *waistcoat* and *trousers* in the USA. Avoid possible ambiguity in your writing, so never write *corn*, even if that is what your textbook uses; specify wheat, maize, *Triticum, Zea*, as appropriate.

6. Revising Written Work

Preceding chapters have dealt with the reading, note-taking, thinking, synthesising and organising that permit you to capture your thoughts and your evidence in a first draft. This chapter concerns the revising that must follow, in which you examine the first draft critically and diagnose and treat the patient as necessary. I (JP) typically revise my own writing four or five times before letting anyone else see it and several more times after it has been reviewed by others. Good writers aren't necessarily more intelligent than bad ones; they just revise more often.

Writing a first draft gives you the opportunity to get facts, ideas and phrasings on paper, where they won't escape. You can then concentrate on reorganising and rephrasing your thoughts in the clearest, most logical way. Revising your work clarifies thinking, improves communication and often gives you a firmer understanding of what you are writing about.

It is difficult to revise your own work effectively unless you can examine it with a fresh eye. After all, you know what you wanted to say; without some distance from the work, you can't really tell whether you've actually said it. For this reason, plan to complete your first draft at least five days before the final product is due, to allow time for careful revision. Reading your work aloud – and listening to yourself as you read – often reveals weaknesses that you would otherwise miss. It also helps to have one or more fellow students read and comment on your draft. It is always easier to identify writing problems – wordiness, ambiguity, faulty logic, faulty organisation, spelling and grammatical errors – in the work of others, so that forming a peer-editing group is a clear step towards more effective writing. Be sure to tell readers of your work that you sincerely want constructive criticism, not a pat on the back.

No matter how sound or brilliant your thoughts and arguments are, it is the manner in which you express them that will determine whether or not they are understood and appreciated (or, in later life, whether they are even read). With pencil or pen at the ready (and scissors and tape, too, if you are not using a computer), edit your first draft for content, clarity, conciseness, flow (coherence), spelling, punctuation and grammar. If you are writing using a word-processor, make your revisions on printed copy rather than on-screen; to edit effectively you must see more than one screen of text at a time. Continue editing and revising – printout by printout – until your work is ready for the lecturer, reviewers, admissions committee or potential employer.

REVISING FOR CONTENT

1. **Make sure every sentence says something.** Consider the following opening sentence for an essay on the tolerance of changes in salinity by estuarine fish:

   > Salinity is a very important factor in marine environments.

   What does this sentence say? What *is* important about salinity? A careful editor will delete the sentence and begin anew with something worth reading. For example,

   > Estuarine fish may be subjected to enormous changes in salinity within only a few hours.

   Similarly, a sentence like

   > There are many physical and biological factors that affect the growth of insect populations.

could profitably be revised to read

> Growth rates of insect populations are influenced by such environmental factors as temperature, food supply, build-up of metabolic wastes, availability of mates and magnitude of predation pressure.

The authors of the revised opening sentences know where their essays are heading, and so does the reader.

2. **Use the word 'relatively' only when making an explicit comparison.** Consider this example: *Many of the animals living near deep-sea hydrothermal vents are relatively large.* Relative to what? Either delete the word and replace it with something of substance (*e.g. Animals living near deep-sea hydrothermal vents can exceed lengths of three metres*) or make a real comparison (*e.g. Many of the animals living near deep-sea hydrothermal vents are large relative to their shallow-water counterparts* or *Some animals living near deep-sea hydrothermal vents are many times larger than their shallow-water counterparts*). In fact, whenever you imply a comparison, make clear what is being compared with what. It is no use writing *Viruses have higher rates of reproduction,* unless one makes clear what other organisms are being compared with viruses for their reproductive rates.

   Another word which must be used accurately is 'literally', which does not mean 'almost'. Only write: *The cattle farmer was literally up to his eyeballs in manure* if the manure really did come right up to his eyes and he was in it.

3. **Do not tell a reader that something is interesting.** Let the reader be the judge. Ask yourself *why* you find it interesting, then make a statement that will interest the reader.

4. **Be cautious in drawing conclusions, but not excessively so.** It is wise to be careful when interpreting biological data, particularly when you have access to only a few experiments or small data sets. For instance, write *These data suggest that . . .* rather than *These data demonstrate, or*

prove that . . . . But don't get carried away, as in the following example:

> This suggests the possibility that inductive interactions between cells may be required for the differentiation of nerve tissue.

Here, the author hedges three times in one sentence, using the words *suggests, possibility,* and *may.* Never hedge more than once per sentence:

> This suggests that inductive interactions are required for the differentiation of nerve tissue.

If you are too unsure of your opinion to write such a sentence, re-examine your opinion.

5. **While revising for content, keep in mind an audience of your peers.** Be sure to define any unfamiliar scientific terms and abbreviations. Giving brief definitions will help keep the attention of readers who may not know the meaning of some terms, and will demonstrate to staff that you know the meaning of the specialised terminology you use. Try to make your writing self-sufficient; the reader should not have to consult textbooks or other sources in order to understand what you are saying.

6. **Watch out for double meanings.** Double meanings involve ambiguity, so should be avoided. This example from an A-level Biology paper is unlikely to cause confusion in the reader, but is included for its humour. *A poplar tree can break wind at a distance of up to 200 metres.*

REVISING FOR CLARITY

Be sure each sentence says what it is supposed to say; you want the reader's head to be nodding up and down, not side to side. Which way is the reader's head going in the following example?

> These methods have different resorption rates and tail shapes.

Do methods have tails? Can methods be resorbed? This sentence fails to communicate what its author had in mind. Here is another sentence that does not reflect the intentions of its author:

> From observations made in aquaria, feeding rates of the fish were highest at night.

How many observers do you suppose can fit into an aquarium? A revised sentence might read,

> Feeding rates of fish held in aquaria were highest at night.

Some biologists are clearly very dedicated to their research:

> Ferguson (1963) examined autoradiographs of sea star digestive tissue after being fed radioactive clams.

Perhaps we should feed the clams not to Ferguson but to the sea stars?

> Ferguson (1963) fed radioactive clams to sea stars and then examined autoradiographs of the sea star digestive tissue.

Note in the above example the advantages of summarising a study in the order in which steps were undertaken; grammatical difficulties typically vanish and the sentence automatically becomes clearer.

Confusing sentences also arise when three or more nouns are lined up in a row. Consider this example:

> Sleep study results show that tryptophan significantly decreases the time needed to fall asleep (Miller and Brown, 1991).

At the first reading, the reader probably expects 'results' to be a verb, but instead it is a noun, preceded by two other nouns. The reader must stop and decode the sentence. The author is discussing the results of studies of people sleeping. We can rewrite the sentence to make this clear.

> Recent studies show that tryptophan decreases the time needed for people to fall asleep (Miller and Brown, 1991).

Think twice before leaving more than two nouns together; two is company, three is a crowd. Here are additional examples of unclear writing:

> This determination was based on mannitol's relative toxicity to sodium chloride.

> The surface area of mammalian small intestines is three to seven times greater than reptiles.

How can one chemical be toxic to another chemical? The author is probably trying to tell us that two chemicals differ in their toxicity to some organisms. With the second example, one wonders how an intestinal surface area can be greater than a reptile; again, the author is not making the comparison he or she intended. It is a non-parallel comparison. In any event, never invoke the 'You know what I mean' defence. If a student writes, 'A normal human foetus has 46 chromosomes', how can one be sure that the student understands that each *cell* of the foetus has 46 chromosomes? It is your job to inform the reader, never the reader's job to guess what you are trying to convey.

Make each sentence state its case unambiguously. Here is a sentence that does not do so:

> Sea stars prey on a wide range of intertidal animals, depending on their size.

Is the author talking about the size of the sea stars that are preying or about the size of the intertidal animals that are preyed upon? Don't be embarrassed at finding sentences like this one in early drafts of your papers and reports. Be embarrassed only if you fail to edit them out.

The Dangers of 'It' and Other Pronouns

Frequent use of the pronouns *it, they, these, their, this* and *them* in your writing should sound an alarm. Probable ambiguity lies ahead, as in the following examples:

> The chemical signal must then be transported to the specific target tissue, but it is effective only if it possesses appropriate receptors.

What are the 'it's'? Are these receptors needed by the chemical signal, or by the target tissue? Next, 'these' causes a problem.

> Antigens encounter lymphocytes in the spleen, tonsils and other secondary lymphoid organs. These then proliferate and differentiate into fully mature, antigen-specific effector cells.

Presumably the lymphocytes are proliferating, not the tonsils, although the author has certainly not made this clear. The problem is easily solved by beginning the second sentence with 'The lymphocytes . . . .'

> Like fanworms and earthworms, leeches have proved very useful to neurophysiologists. Their neurons are few and large, making them particularly easy to study with electrodes.

Who or what are 'Their'? Readers may be surprised to learn that neurophysiologists have so few neurons and are so easy to study!

> Tropical countries are home to both venomous and nonvenomous snakes. They kill their prey by constriction or by biting and swallowing them.

What are 'they' and 'them'? How much clearer the last sentence could become by replacing *they* with a few words of substance and by deleting *them* entirely:

> Tropical countries are home to both venomous and nonvenomous snakes. The nonvenomous snakes kill their prey by constriction or by biting and swallowing.

Finally, look what can happen when a variety of these pronouns are scattered throughout a sentence:

> Although they both saw the same things in their observations of embryonic development, they had different theories about how this came about.

A patient reader of the whole essay could probably work out this sentence eventually, but its author has certainly violated Rule 7 (page 6), 'Never make the reader turn back', in a most extreme fashion.

When editing your work, read it carefully and with scepticism, checking that you have written exactly what you mean. Never make the reader guess what you have in mind. Everything you write must make sense – to yourself and to the reader. As you read each sentence you have written, think: what does this sentence say? What did I mean it to say?

You need not be an expert grammarian to write correctly and clearly. You should study Chapter 5 carefully, but with practice, especially if you read your work aloud, you can often recognise a sentence in difficulty and sense how to correct it without knowing the name of the grammatical rule that has been violated.

REVISING FOR COMPLETENESS

Make sure each thought is complete. Be specific in making assertions. The following statement is much too vague:

> Many insect species have been described.

How many is 'many'? After editing, the sentence might read,

> Nearly one million insect species have been described.

Similarly, the following sentence

> More caterpillars chose diet A than diet B when given a choice of the two diets (Fig. 2).

would benefit from this alteration, from the author thinking quantitatively:

> Nearly five times as many caterpillars chose diet A than diet B when given a choice of the two diets (Fig. 2).

Here is another kind of incompleteness:

> If diffusion was entirely responsible for glucose transport, then this would not have occurred.

This rears its ugly head again, as the author avoids the responsibility of drawing a clear conclusion and forces the reader to go back. Even the beginning of the sentence is unnecessarily vague since, it turns out, the discussion is concerned only with glucose transport in intestinal tissue. Try to make your sentences tell a more detailed story, as in this revision:

> If diffusion were entirely responsible for glucose transport into cells of the intestinal epithelium, transport would have continued when the inhibitors were added.

Be especially careful to revise for completeness whenever you find that you have written *etc.* In writing a first draft, use *etc.*'s freely when you'd rather not interrupt the flow of your thoughts by thinking about exactly what 'other things' you have in mind. When revising, replace most, but not necessarily all, *etc.*'s with words of substance. You should find yourself thinking, 'What exactly do I have in mind here?' If you come up with additional items for your list, add them unless there are too many to include. If you find that you have nothing to add, simply replace the *etc.* with a full stop. Consider the following sentence and its two improvements:

Plant growth is influenced by a variety of environmental factors, such as light intensity, nutrient availability, etc.

Revision 1: Plant growth is influenced by a variety of environmental factors, such as light intensity, day length, nutrient availability and temperature.

Revision 2: Plant growth is influenced by such environmental factors as light intensity, day length, nutrient availability and temperature.

The original version, although grammatically correct, is incomplete, waiting for the reader to fill in the missing information.

REVISING FOR CONCISENESS

Omitting unnecessary words will make your thoughts clearer and more convincing. In particular, such phrases as, 'It should be noted that', 'It is interesting to note that', and 'The fact of the matter is that' are common in first drafts, but should be ruthlessly eliminated in preparing the second. Such verbal excess also takes less conspicuous forms. How might you shorten this next sentence?

Dr Smith's research investigated the effect of pesticides on the reproductive biology of birds.

Who did the work: Dr Smith or his research? A reasonable revision would be:

Dr Smith investigated the effect of pesticides on the reproductive biology of birds.

We have eliminated one word and the sentence has not suffered.

Working further, we could replace 'the reproductive biology of birds' with 'avian reproduction', achieving a reduction of three more words:

The next example requires similar attention:

> It was found that the shell lengths of live snails tended to be larger for individuals collected closer to the low tide mark (Fig. 1).

A good editor would eliminate the first phrase of that sentence and prune further from there. In particular, what does the author mean by 'tended to be larger'? Here is an improved version:

> Live snails collected near the low tide mark had greater average shell lengths (Fig. 1).

Most wordy sentences suffer from one or several of four major ailments and can be restored to health by obeying the Four Commandments of Concise Writing.

First Commandment: Eliminate Unnecessary Prepositions

Consider this example:

> The results indicated a role of haemal tissue in moving nutritive substances to the gonads of the animal.

Any sentence containing such a long string of prepositional phrases – *of* tissue, *in* moving substances, *to* the gonads, *of* the animal – is a candidate for the editor's operating table. This sentence actually contains a simple thought, buried amidst a clutter of unnecessary words. After surgery, the thought emerges clearly:

> The results indicated that haemal tissue moved nutrients to the animal's gonads.

Here is another example:

> The cells respond to foreign proteins by rapidly dividing and starting to produce antibodies reactive to the protein groups that induced their production.

The reader's head spins, an effect avoided by this more concise version:

> In the presence of foreign proteins, the cells divide rapidly and produce antibodies against those proteins.

By eliminating prepositions, *Gould arrives at the conclusion that . . .* becomes *Gould concludes that. . . . Grazing may constitute a benefit to . . .* becomes *Grazing may benefit. . . . These data appear to be in support of the hypothesis that . . .* becomes *These data appear to support the hypothesis that . . .*, and *Schooling of fish is a well documented phenomenon* becomes *Fish schooling is well documented.*

Second Commandment: Avoid Weak Verbs

Formal scientific writing can be boring when the sentences contain no real action; commonly, the colourless verb *to be* is used where a more vivid verb would be more effective, as in this example:

> The fidelity of DNA replication is dependent on the fact that DNA is a double-stranded polymer held together by weak chemical interactions between the nucleotides on opposite DNA strands.

This patient suffers from Wimpy Verb Syndrome, with a slight touch of Excess Prepositional Phrase. There is potential action in this sentence, but it is sound asleep in the verb phrase 'is dependent'. Converting to the stronger verb 'depends', we read,

> The fidelity of DNA replication depends on the fact that DNA is a double-stranded polymer. . . .

But why stop there? Let's eliminate some clutter (*on the fact that*) and another weak verb ('is'):

> The fidelity of DNA replication depends on DNA being a double-stranded polymer. . . .

Similarly,

> Plant vascular tissues function in the transport of food through xylem and phloem.

can be enlivened by converting the phrase *function in the transport of* to the more vigorous verb *transport*:

> Plant vascular tissues transport food through xylem and phloem.

By choosing a stronger verb, we have also eliminated two prepositional phrases (*in the transport of* and *of food*). Step by step, the sentence becomes shorter and clearer. As often happens during revision, fixing one problem reveals an additional problem, in this case a fundamental structural weakness that makes the reader wonder whether the student understands the relationship between *plant vascular tissues* and *xylem and phloem*. Revising for content, we might rewrite the sentence as

> Plants transport nutrients through their vascular tissues, the xylem and phloem.

Third Commandment: Do Not Overuse the Passive Voice

The passive voice is often a great enemy of concise writing, in part because the associated verbs are weak. If the subject (rats and mice, in the following example) is on the receiving end of the action, the voice is passive:

> Rats and mice were experimented on by him.

If, on the other hand, the subject of a sentence ('He', in the coming example) is on the delivering end of the action, the voice is active:

He experimented with rats and mice.

Note that the 'active' sentence contains only six words, while its 'passive' counterpart contains eight. In addition to creating excessively wordy sentences, the passive voice often makes the active agent anonymous, and a weaker, sometimes ambiguous sentence may result:

Once every month for two years, mussels were collected from five intertidal sites in Barnstable County, MA.

Whom should the reader contact if there is a question about where the mussels were collected? Were the mussels collected by the writer, by fellow students, by a lecturer, or by a private company? Eliminating the passive voice clarifies the procedure:

Once every month for two years, members of the class collected mussels from five intertidal sites in Barnstable County, MA.

Similarly, *It was found that* becomes *I found*, or *we found*, or, perhaps, *Smith (1986) found*. Whenever it is helpful for the reader to know who the agent of the action is, and whenever the passive voice makes a sentence unnecessarily wordy, use the active voice:

Passive: Little is known of the nutritional requirements of these animals.

Active: We know little about the nutritional requirements of these animals.

Passive: The results were interpreted as indicative of . . . .

Active: The results indicated . . . .

Fourth Commandment: Make the Organism the Agent of the Action

Studies on the rat show that the activity levels vary predictably during the day (Hatter, 1976).

This is not a terrible sentence but it can be improv
action from the studies (*Studies show*) to the organism ir

> Rats vary their activity levels predictably during th
> (Hatter, 1976).

The revised sentence is shorter, clearer and more interesting because now an organism is *doing* something. Be a person of few words; your readers will be grateful.

REVISING FOR FLOW

A strong paragraph – indeed, a strong paper – takes the reader smoothly and inevitably from a point upstream to one downstream. Link your sentences and paragraphs using appropriate transitions, so that the reader moves effortlessly and inevitably from one thought to the next, logically and unambiguously. Remind the reader of what has come before and help the reader anticipate what is coming next. Consider the following example:

> Since aquatic organisms are in no danger of drying out, gas exchange can occur across the general body surface. The body walls of aquatic invertebrates are generally thin and water permeable. Terrestrial species that rely on simple diffusion of gases through unspecialised body surfaces must have some means of maintaining a moist body surface, or must have an impermeable outer body surface to prevent dehydration; gas exchange must occur through specialised, internal respiratory structures.

This example gives the reader a choppy ride and cries out for careful revision of the way the ideas are presented. In the following revision, note the effect of two important transitional expressions, *thus* and *in contrast to*. The first connects two thoughts, while the second warns the reader of an approaching shift in direction.

> Since aquatic organisms are in no danger of drying out, gas exchange can occur across the general body surface. Thus, the body walls of aquatic invertebrates are generally thin and water-permeable, facilitating such gas exchange. In contrast to the simplicity of gas exchange mechanisms among aquatic species, terrestrial species that rely on simple diffusion of gases through unspecialised body surfaces must either have some means of maintaining a moist body surface, or must have an impermeable outer body covering that prevents dehydration. If the outer body wall is impermeable to water and gases, respiratory structures must be specialised and internal.

In the first draft, the reader must struggle to find the connection between sentences. In the revised version, the writer has assisted the reader by connecting the thoughts, resulting in a more coherent paragraph. Here is another example of a stagnating paragraph:

> The energy needs of a resting sea otter are three times those of terrestrial animals of comparable size. The sea otter must eat about 25% of its body weight daily. Sea otters feed at night as well as during the day.

Revising for improved flow produces the following. Note that the writer has introduced no new ideas. The additions, here underlined, are simply clarifications that make the connections between each point explicit.

> The energy needs of a resting sea otter are three times those of terrestrial animals of comparable size. <u>To support such a high metabolic rate</u>, the sea otter must eat about 25% of its body weight daily. <u>Moreover</u>, sea otters feed continually, at night as well as during the day.

The following transitional words and phrases are especially useful in linking thoughts to improve flow: *in contrast, however, although, thus, whereas, even so, nevertheless, moreover, despite the, in addition to.*

Repetition and summary are also highly effective ways to link thoughts. For instance, repetition has been used to connect the first two sentences of the revised example about sea otters: 'To support such a high metabolic rate' essentially repeats, in summary form, the information content of the first sentence. Repetition is a particularly effective way of linking paragraphs; in reminding the reader of what has come before, the author consolidates his or her position and then moves on.

Judicious use of the semicolon and colon (see Chapter 5) can also ease the reader's journey. In particular, when the second sentence of a pair explains or clarifies something contained in the first, you may wish to combine the two sentences into one with a semicolon. Consider the following two sentences:

> This enlarged and modified bone, with its associated muscles, serves as a useful adaptation for the panda. With its 'thumb', the panda can easily strip the bamboo on which it feeds.

The reader probably has to pause to consider the connection between the two sentences. Using a semicolon, the passage would read:

> This enlarged and modified bone, with its associated muscles, serves as a useful adaptation for the panda; with its 'thumb', the panda can easily strip the bamboo on which it feeds.

The semicolon links the two sentences and eliminates an obstruction in the reader's path. Similarly, a semicolon provides an effective connection between thoughts in the following example:

> Recombinant DNA technology enables large-scale production of particular gene products; specific genes are transferred to rapidly dividing host organisms (yeast or bacteria), which then transcribe and translate the introduced genetic templates.

REVISING FOR TELEOLOGY AND ANTHROPOMORPHISM

Remember, organisms do not act or evolve with intent (page 9). Consider the following examples of teleological writing and learn to recognise the trend in your own work:

> Barnacles are incapable of moving from place to place, and therefore had to evolve a specialised food-collecting apparatus in order to survive.

> Squid and most other cephalopods lost their external shells in order to swim faster, and so better compete with fish.

> Aggression is a directed behaviour that many sea anemones exhibit to promote the survival of an individual's own genotype.

Revise all teleology out of your writing. Also beware of anthropomorphising, in which you give human characteristics to nonhuman entities, as in this example:

> The existence of sage in the harsh climate of the American plains results from Nature's timeless experimentation.

This conveys a fuzzy picture of how natural selection operates. Beware also of using phrases such as:

> The bacteria preferred the richer medium. . . .

when bacteria are incapable of having conscious preferences, and merely grew better on the richer medium.

REVISING FOR SPELLING ERRORS

Misspellings convey the impression of carelessness, laziness or even stupidity. These are not advisable images to present to staff, prospective employers, or the admissions officers of graduate or professional programmes. Using a spelling-checker computer program will save you from misspelling many non-technical words, but it won't catch such spelling errors as 'is' versus 'if', or 'nothing' versus 'noting', and it is unlikely to be of much help in screening technical terms for you. Use the computer for a 'first pass', but use your own eyes for the second.

Check the list of frequently misspelled words given in Chapter 5, pages 169–70, and the hints there on spelling. Note also that *mucus* is a noun; as an adjective, the same slime becomes *mucous*. Many animals produce mucus, and mucous trails are produced by snails.

REVISING FOR GRAMMAR AND PROPER WORD USAGE

Appendix B (pages 249–51) lists a number of books that include excellent sections on grammar and word usage, and see Chapter 5 here for advice. While on the look-out for incomplete sentences, run-on sentences, faulty use of commas, faulty comparisons, disagreements between subjects and verbs, and other grammatical blunders, you should also look out for the following troublesome points:

- *its/it's.* The possessive pronoun is *its*:

  When treated with the chemical, the protozoan lost its cilia and died.

  It's is always an abbreviated form of *it is* or *it has*:

  It's clear that the chemical treatment caused the loss of cilia.

In general, contractions are not welcome in formal scientific writing. Thus you can avoid the problem by writing *It is* when appropriate, not *It's*.

- ***i.e./e.g.*** These abbreviations are not interchangeable. '*I.e.*' is an abbreviation for *id est*, which in Latin means *that is* or *that is to say*. For example:

  In mammals, there are two sexes and two normal sex karyotypes, i.e. female (XX) and male (XY).

  In contrast, '*e.g.*' stands for *exempli gratia*, meaning 'for example'.
  Here, '*e.g.*' is used to indicate that what follows is only a partial listing of references.

  However, the larvae of several butterfly species (e.g. Papilio demodocus Esper, P. eurymedon and Pieris napi) are able to feed and grow on plants that the adults never lay eggs on.

- ***However, therefore, moreover.*** These words are often incorrectly used as conjunctions, as in the following examples:

  The brain of a toothed whale is larger than the human brain, however the ratio of brain to body weight is greater in humans (Table 4).

  The resistance of mosquito fish to the pesticide DDT persisted into the next generation, therefore the resistance was probably genetically based.

  Protein synthesis in frog eggs will take place even if the

nucleus is surgically removed, moreover the pattern of protein synthesis in such enucleated eggs is apparently normal.

- These examples all demonstrate the infamous comma splice, in which a comma is mistakenly used to join what are really two separate sentences (see Chapter 5, page 148). Reading aloud, you should hear the material come to a complete stop before the words 'however', 'therefore', and 'moreover'. Thus you must replace the commas with either a semicolon or a full stop, as in these revisions of the first example:

    The brain of a toothed whale is larger than the human brain; however, the ratio of brain to body weight is greater in humans.

    The brain of a toothed whale is larger than the human brain. However, the ratio of brain to body weight is greater in humans.

  And don't forget: ***The data are*** . . . (see page 174).

BECOMING A GOOD EDITOR

The best way to become an effective reviser of your own writing is to become a critical reader of other people's writing. When you read a newspaper, magazine, or textbook, be on the look-out for ambiguity, errors and wordiness, and think about how the passage might best be rewritten. You will gradually come to recognise the same problems, and the solutions to these problems, in your own writing. But don't try to fix everything at once. Whether you are editing an early draft of your own work or a fellow student's work, be concerned first with content. Until you are convinced that the author has something to say, it makes little sense to be greatly concerned with how he or she has said it. Always leave at least several days to make revisions.

Take an especially careful look at the title and the first few paragraphs. Does the title indicate exactly what the paper or lab report is about? Do the title and first paragraph seem closely related? Does one sentence lead logically to the next, establishing a clear direction for what follows? Can you tell from the first paragraphs exactly what this paper, proposal, or report is about, and why the issue is of interest? Or are you reading a series of apparently unrelated facts that seem to lead nowhere, or in many different directions?

Second drafts commonly arise from only a small portion of the first – perhaps a few sentences buried in the last third of the original. In such a case you must abandon most of the first draft and begin afresh, but this time you are writing from a stronger base.

Once the piece has a clear direction you can revise for flow and clarity. Does each sentence make sense, and does each sentence and paragraph lead in a logical fashion to the next? Does the concluding paragraph address the issue posed in the first paragraph?

If examining a lab report, study the Results section first. Does it conform to the requirements outlined in Chapter 3? Does the Materials and Methods section answer all procedural questions that were not addressed in the figure captions and table legends? Should some of those questions (*e.g.* experimental temperature) be addressed in the captions and legends? Does the Introduction state a clear question and provide the background information needed to understand why that question is worth asking? Does the Discussion interpret the data and clearly address the specific issue raised in the Introduction? Only when you can answer 'yes' to these questions should you worry about fine-tuning the paper, editing for conciseness, completeness, grammar and spelling.

Giving Criticism

Look at someone else's paper in the same way that you should look at your own, concerning yourself first with content. Comments on spelling and grammar are more appropriate to later drafts than to the first draft. When examining a first draft, it may be most useful to write the author a few paragraphs of commentary and not write on the paper at all. Don't rewrite the paper for the author; your role is to point out strengths and perceived weaknesses and to offer the best advice you can. Here is an example of a student making comments about the first draft of a research proposal written by a fellow student.

David, I think you have a good idea for a project here, but it's not reflected in your introduction or the title. The question you finally state in the middle of p.4 caught me completely by surprise; at least until the bottom of p.2 I thought you were interested in the effects of electromagnetic fields on human development, and by the end of p.3, I wasn't sure what you were planning to study! On pp.2-3 especially, I couldn't see how the indicated paragraphs (see my comments on your paper) related to the question you ended up asking. Or perhaps they are relevant, but you haven't made the connections clear? The entire introduction seems to be in the 'book report' format we discussed in class, rather than a piece of writing with a point to make (I'm having this trouble, too). The information you present is interesting but a lot of it seems irrelevant. Try to make clearer connections between the paragraphs. Here is a possible reorganisation plan: introduce the concept of electromagnetic fields in the first few sentences (what they are, what produces them); then mention potential damaging effects on physiology and development (it's not clear why the question is so important until one gets to p.6!); then state your question and note why urchins are especially good animals to study. Will that work?

In the Introduction, I would expand the paragraph on gene expression effects; discuss one or two of the key experiments in some detail, rather than just tell us the results. I think this is important, since your experiments are a follow-up on these.

Your experimental design seems sound, although I'm not sure the experiments really address the exact question you pose in your introduction (see my comments on the draft; probably you just need to rephrase the question). But I didn't see any mention of a control; without the control, how will you be sure that any effects you see are due to the electromagnetic field? Also, won't your treatment raise the water temperature? If so, you will need to control for that as well.

> Finally, you might want to ask Dr Lane about this, but I think you should write for a more scientifically advanced audience. Your tone seems a bit too chatty and informal. And watch those prepositions – you use them almost as freely as I do! I enjoyed reading your paper and look forward to seeing the next draft!

Notice that this reviewer points out the strengths of the piece without overlooking the weaknesses and deals with the major problems first. Be firm but kind in your criticism; your goal is to help your colleague, not to crush his or her ego. Be especially careful to avoid sarcasm. Write a page of constructive criticism that you would feel comfortable receiving.

Receiving Criticism

Be pleased to receive suggestions for improving your work. A colleague who returns your paper with only a smile and a pat on the back does you no favours. It is good to receive *some* positive feedback, of course, but what you should really hope for is constructive criticism. On the other hand, don't feel you must accept every suggestion offered. Examine each one honestly and decide for yourself if the reader is on target or not; these reviews are advisory only, giving you a chance to see how another person interprets what you have written. Sometimes the reader will misinterpret your writing and you may therefore disagree with the specific criticisms and suggestions; but something unclear to one reader may be unclear to others. Try to work out where the reader went astray, and modify your writing to prevent future readers from following the same path.

It is hard to read criticism of your writing without feeling defensive, but learning to value those comments puts you firmly on the path to becoming a more effective writer. If you are not communicating well, you need to know it, and to know why.

Fine-tuning

Once the writing has a clear direction and solid logic, it is time to make one or two final passes to see that each sentence is doing its job in the clearest, most concise fashion. As a first step in developing your ability to fine-tune writing, read the following sentences and try to put in words the ailment afflicting each one. Then revise those sentences that need help. Pencil your suggested changes directly onto the sentences, using the guide to proof-readers' notation presented in Table 5 and the following example:

> Hermaphroditism is Commonly encountered among invertebrates. For example, the young East Coast oyster, *Crassostrea virginica*, matures as a male, later decomes a female and may chenge sex every few years there after. sequential hermaphrodites generally change sex only once, and usually change from male to female. In contraast to species than change sex as they age, many invertebrates are simultaneous hermaphrodites. Self-fertilization is rare among simultaneous hermaphrodites it can occur, as in the tapeworms.

It is wise when editing someone else's work to use a different coloured pen or pencil, to be sure the reader will see suggested changes.

Table 5. Proof-reader's symbols used in revising copy.

Problem	Symbol	Example
1. Word has been omitted	^ (caret)	study describes ^ effect (the)
2. Letter has been omitted	^ (caret)	that bo^k (o)
3. Letters are transposed	(transposition mark)	form the sea
4. Words are transposed	(transposition mark)	was only exposed
5. Letter should be capitalised	≡ (three short underlines)	these data
6. Letter should be lower-case	/ (slash)	These Data
7. Word should be in italics	(underline once)	Homo sapiens
8. Words are run together	| (draw vertical line between)	edit|carefully
9. Word should be deleted	(draw line through)	the ~~nice~~ data
10. Space should not have been left	(sideways parentheses)	the e nd
11. Wrong letter	/ (draw line through and add correct letter above)	hemale (f)
12. Wrong word	(draw line through and add correct word above)	~~This~~ data (These)
13. Need to begin a new paragraph	¶ (paragraph symbol)	female. ¶ In contrast
14. Restore original	STET	The ~~energy~~ needs (STET)

Sentences in Need of Revision

1. To perform this experiment there had to be a low tide. We conducted the study at Blissful Beach on September 23, 1991, at 2:30 PM.
2. In *Chlamydomonas reinhardi*, a single-celled green algae, there are two matine types, + and –. The + and – cells mate with each other when starved of nitrogen and form a zygote.
3. Protruding form this carapace is the head, bearing a large pair of second antennae.
4. The order in which we think of things to write down is rarely the order we use when we explain what we did to a reader.
5. The purpose of Professor Wilson's book is the examination of questions of evolutionary significance.
6. Swimming in fish has been carefully studied in only a few species.
7. One example of this capcity is observed in the phenomenon of encystment exhibited by many fresh water and parasitic species.
8. An estuary is a body of water nearly surrounded by land whose salinity is influenced by freshwater drainage.
9. In textbooks and many lectures, you are being presented with facts and interpretations.
10. The human genome contains at least 50,000 genes, however there is enough DNA in the genome to form nearly 2×10^6 genes.
11. These experiments were conducted to test whether the condition of the biological films on the substratum surface triggered settlement of the larvae.
12. Various species of sea anemones live throughout the world.
13. This data clearly demonstrates that growth rates vary with temperature.
14. Hibernating mammals mate early in the spring so that their offspring can reach adulthood before the beginning of the next winter.
15. This study pertains to the investigation of the effect of this pesticide on the orientation behavior of honey bees.
16. The results reported here have led the author to the conclusion that thirsty flies will show a positive response to all solutions, regardless of sugar concentration (see Figure 2).
17. Numbers are difficult for listeners to keep track of when they are floating around in the air.

18. Measurements of respiration by the salamanders typically took one-half hour each.
19. The results suggest that some local enhancement of pathogen specific antibody production at the infection site exists.
20. Usually it has been found that higher temperatures (30°C) have resulted in the production of females, while lower temperatures (22–27°C) have resulted in the production of males. (e.g., Bull, 1980; Mrosousky, 1982)

There are several ways to improve each of the preceding sentences. Possible revisions are shown in Appendix C (pages 253–4), but you should make your own modifications before looking at mine (JP). Be sure that you can identify the problem suffered by each original sentence, that you understand how that problem was solved by my revision, and that your revision also solves the problem (and does not introduce any new difficulties).

7. Answering Examination Questions

Well before the exam, look at copies of past papers if they are available, to see the format and the kinds of questions asked. Try a few sample questions. Start your revision long before you think it is really necessary, making sure that your lecture notes are in a good form for revising. Prepare summaries of key facts and test yourself on them. For the actual exam, remember to take pens, with spares, pens with different colours for diagrams, ruler, pencils, rubber, white correcting fluid, calculator (if permitted) with spare batteries, and any medicines you might need, especially if exams give you headaches or asthma attacks.

ALLOCATING YOUR TIME

When you look at the paper, see how long you have got and how many questions you have to answer. Be absolutely sure to put your name on every answer book. If there is a choice of questions, decide which you will do and start on the one you like most. Do not, however, spend too much time on that question, at the expense of others. Spend almost exactly equal amounts of time on each question, or if different questions carry different numbers of marks, spend your time in proportion to the marks available. The first 20 per cent of the marks on any question are the easiest to get and the last 20 per cent are the hardest. A student who has to answer four questions, and who spends half the time on his or her favourite question, is being foolish as there is then not enough time to answer the three remaining questions properly. Answering only three questions instead of four is also a big handicap, as you will then need to average 67 per cent on each question even to get 50 per cent as your final mark.

Suppose you have 45 minutes for a question which is partly a numerical problem and partly a related descriptive section. If you spend 20 minutes on the numerical answers, then spend 25 minutes on the descriptive section. When doing numerical problems, show as much of your reasoning and workings as possible, so that you can get credit for the right approach, even if your final numerical answer is wrong. Be sure to give any units for the numbers.

The major faults in undergraduate answers, apart from lack of knowledge and understanding, are wrong distribution of time on different questions, not answering the right number of questions, and answering the question the student hoped for instead of the actual question set.

ANSWERING THE QUESTION SET

For all types of question, read each question slowly at least three times, trying to work out exactly what is required. Students sometimes see a relevant word in a question and unthinkingly regurgitate that part of their lecture notes, irrespective of the actual question asked. The real question may well need you to integrate material from several different lectures. With some questions, you can even use material from different courses if relevant.

Give named examples wherever possible. This creates an impression of real knowledge. Do use well-labelled diagrams to convey information, using different colours where that helps. If you use graphs, label both axes. Do not waste time by copying out the questions, so long as it is clear which part of which question you are answering. There is no point in starting with a phrase such as 'This essay is all about. . . .'

ANSWERING DIFFERENT TYPES OF QUESTION

For multiple-choice questions, you need to think carefully and to have detailed factual knowledge of the course. Read the question and the list of possible answers at least twice. If you are sure of the answer, put a tick or cross as instructed. What you do if you are unsure depends on how the exam is marked, which you should ask staff about well before the exam. In many such exams, marks are deducted for incorrect answers, in which case it is usually better to leave an answer blank than to have a random guess at questions to which you do not know the answer. If you are reasonably sure of an answer, then put it down. Sometimes you may not know the correct answer, but you may be able to deduce that all the listed possibilities except one are incorrect.

With short-answer questions, answer in complete sentences, giving the essence of the answer: you will not have time to add a lot of details, but do include named examples if possible. For essay questions, use sub-headings to give your essay more structure. At the end, round the essay off with a sentence or two of summary; don't just end abruptly. Never waste time on throw-away comments such as 'Sorry, I didn't revise this', or 'Ran out of time!' It is your responsibility to organise your time properly.

Apart from limitations of time, answering essay questions differs little from other forms of scientific writing already discussed. Follow all the guide-lines outlined in Chapter 1. Don't forget, you must read the question carefully before writing anything. You must answer the question posed, not the question you would have preferred to see. Note whether the question asks you to list, discuss, or compare. A list will not satisfy the requirements of a discussion or comparison. A request for a list tests whether you know all components of the answer; a request for a discussion additionally examines your understanding of the interrelationships among these components.

Consider this list of the characteristics of a Big Mac and a Whopper, based on a study conducted in Massachusetts in 1991.

Whopper	**Big Mac**
1 beef patty	2 beef patties
patty 3.75–4" diameter	patties 3.25" diameter
broiled beef	fried beef
2-part bun (top & bottom)	3-part bun (3 slices)

Whopper	**Big Mac**
sesame seeds on top bun	sesame seeds on top bun
slice of pickle	slice of pickle
slices of onion	chopped onion
2–3 slices of tomato	slice of cheese
ketchup	lettuce
mayonnaise	sauce
$1.99	$1.99
packed in a cardboard box	surrounded by a cardboard ring and wrapped in paper

Suppose you are asked to write an essay presenting the features of both items. Your essay might look like this:

> The Big Mac consists of two patties of fried ground beef, each patty approximately 3.25 inches in diameter, with lettuce, chopped onion, sliced pickle, a slice of cheese, some reddish sauce, and a three-part bun, with the two patties separated from each other by one of the slices of bun. The top slice of the bun is covered with sesame seeds. The Big Mac sells for $1.99 and is served in a paper wrapper, with a cardboard ring inside to hold the sandwich together.
>
> The Whopper consists of one slice of broiled ground beef (approximately four inches in diameter), with mayonnaise, ketchup, several slices each of tomato, pickle, and onion, and a two-part bun, with the upper half of the bun covered with sesame seeds. The Whopper sells for $1.99 and is served in a cardboard box.

If you are asked to compare, or to compare and contrast, the two products, your essay must be written differently:

> Both the Big Mac and the Whopper contain ground beef and are served on buns. The two hamburgers differ, however, with regard to the way the meat is cooked, the way the meat and bread are distributed within the hamburger, the nature

> of accompanying condiments, and how the sandwiches are served.
>
> The meat in the Big Mac is fried, and each sandwich contains two patties, each approximately 3.25 inches in diameter and separated from the second patty by a slice of bun. In contrast, the meat in the Whopper is broiled, and each sandwich contains a single, larger patty, approximately 3.75–4 inches in diameter. The top bun of both sandwiches is dotted with sesame seeds. Both the Big Mac and the Whopper contain lettuce, onion, and slices of pickle. The Big Mac, however, contains chopped onion, whereas the onion in the Whopper is sliced. Moreover, the Big Mac has a slice of cheese, which is absent from the Whopper. On the other hand, the Whopper comes with slices of tomato, which are absent from the Big Mac. Both sandwiches contain a sauce: ketchup and mayonnaise in the Whopper and a premixed sauce in the Big Mac. The Big Mac and the Whopper both cost $1.99.

If you are asked for a comparison and you discuss the two elements separately, you will lose marks because you have failed to demonstrate your understanding of the relationship between the two products. The facts included are the same in the two essays above: the difference lies in the way the facts are presented.

When asked for a list, give a list; this requires less time than a discussion. When asked for a discussion, discuss. Present the facts and support them with specific examples. When asked for a comparison, you will generally discuss similarities and differences, but the word *compare* can also mean that you should consider only similarities. Often a lecturer will ask you to compare and contrast, avoiding any such ambiguity.

As far as time permits, present all relevant facts. Although there are many ways to answer an essay question correctly, your lecturer will undoubtedly have in mind a series of facts that he or she would like to see included. That is, the ideal answer will contain a finite number of components; the way you deal with each of these components is up to you, but each component should be in your answer. Before you write your essay, list in rough at the top of your answer all components of the ideal answer, drawing from lecture material, practicals, field trips and any further reading you have done. Suppose you are asked the following question: 'Discuss the influence of physical and biological factors on the distribution of plants in a forest.'

What components will the perfect answer to this question contain? List all relevant factors as they occur to you – don't worry about the order, yet:

Physical	**Biological**
amount of rainfall	competition with other plants
annual temperature range	predation by herbivores
light intensity	diseases caused by microbes
hours of light per day	
type of soil	
pesticide use	
nutrient availability	

This list is not your answer to the question; it is an organising vehicle for your use. Abbreviate if you wish ('nutr. avail.', 'pred. by herbs'). Now arrange the elements of your list in a logical order, perhaps from most to least important or so that related elements are considered together. This grouping and ordering are most quickly done by numbering the items in your list in the order that you decide to consider them. Avoid spending all your time discussing a few of these components to the exclusion of others. If you discuss only four of the ten relevant issues, your lecturer may assume that you don't realise that the other issues are also relevant. Lightly cross through each item on your list as you finish writing about it.

As you write, stick to the facts. An examination is not the place to express personal, unsubstantiated opinion. Your lecturer wishes to discover what you have learned and what you understand. Focus, therefore, on facts and support statements of fact or opinion with evidence or examples. You may wish to suggest a hypothesis as part of your essay; if so, be sure to include the evidence upon which your hypothesis is based. Keep the question in mind as you write. Don't include superfluous information. If what you write is irrelevant to the question, you won't get credit for it. Listing the components of your answer before you write your essay will help keep you on track.

Ideally, read through each answer once at the end. You may discover obvious errors, such as omitting the word 'not' from a particular sentence, or forgetting to fill in an area where you whitened out an error and left the fluid to dry before putting in the correct word.

Do remember that someone will have to read and mark your answers, so clarity and legibility are crucial. The lecturer probably has many tasks to do and many scripts to mark in a short time. Illegible passages should receive no marks. Having even one illegible word in a sentence can make its meaning or correctness uncertain. Passages where many words need deciphering will slow up and irritate the marker, who will become increasingly alienated and less likely to give any 'benefit of the doubt'. Although marks are not specifically given for handwriting or clear English, they will influence the marker. Don't forget that many errors of English – such as confusing *affect* and *effect*, or where punctuation errors change the intended meaning – result in very bad science, and the bad science will be penalised for being bad science.

8. Writing Summaries and Critiques

For writing summaries and critiques, you are asked to read a paper from the original scientific literature and summarise or assess it, usually in only one or two pages. *Brief* does not, in this case, mean *easy.* Producing that one- or two-page summary or critique will probably require as much mental effort as an essay five to ten pages long. You must fully understand what you have read, which means reading the paper several times, slowly and thoughtfully. Follow the same procedures whether you are asked to write a summary or a critique. A critique begins as a summary, to which you then add your own evaluation of the paper.

To begin, read the paper once or twice without taking notes, following the advice in Chapter 2. Fight the temptation to underline, highlight, or otherwise create the illusion that you are accomplishing something. It is often difficult to distinguish the significant from the not-so-significant points during the first reading of a scientific paper; skim the paper once for general orientation and overview. Don't try for detailed understanding in the first reading, but do jot down unfamiliar terms or the names of unfamiliar techniques so that you can look these up in a textbook before you reread the paper. It may help to consult a textbook about the biology of the organisms studied.

During the next, more careful reading of the paper, pay special attention to the Materials and Methods and the Results sections; the essence of any scientific paper is contained here. The results obtained depend on the way the study was conducted. Were samples taken only at one particular time of year? Was the study replicated? How many individuals were examined? What techniques were used? In an experiment, what variables (for example, photoperiod, temperature, salinity, or food supply) were held constant? Were proper controls provided for each experiment? Which factors might affect the outcome of the study?

As you begin to study the Results section, scrutinise every graph, table, and illustration, developing your own interpretations of the data before rereading the author's account. We are readily influenced by the opinions of others, especially when those opinions are well-written. Keep an open mind when reading the author's words, but try to form your own opinions about the data first; you may see something that the author did not.

THE FIRST DRAFT

You are ready to write your first draft when you can distil the essence of the paper into a single, intoxicating summary sentence, or, at most, two summary sentences. These sentences should include *all* the key points, present an *accurate* summary of the study, and be *fully comprehensible* to someone who has never read the original paper. As a general rule, do not begin to write your review until you can write such an abbreviated summary; this exercise will help you discriminate between the essential points of the paper and the extra, complementary details.

Once your summary sentence is committed to paper, ask yourself these questions:

1. Why was the study undertaken? To answer this, draw especially from information given in the Introduction and Discussion sections.
2. What specific questions were addressed? Summarise each in a single sentence.
3. What specific approaches were taken to address each question on your list?
4. What were the major findings of the study?
5. What questions remain unanswered by the study? These may be questions addressed by the study but not answered conclusively, or they may be new questions arising from the findings of the study.

WRITING THE SUMMARY

When you can answer these questions without referring to the paper you have read, you can begin to write. Give the complete citation for the paper being discussed: names of all authors, year of publication, title of the paper and of the journal, and volume and page numbers of the article. Begin your summary with a few sentences of background information. Your introductory sentences must lead up to a statement of the specific questions the researchers set out to answer. Next, tell (i) what approaches were used to investigate each question and (ii) what major results were obtained. Be sure to state, as succinctly as possible, exactly what was learned from the study.

To cover so much ground within the limits of one page is no small feat, but it can be done if you first make certain that you fully understand what you have read. Consider the following example of a brief, successful summary. Before writing the summary the student condensed the paper into these two sentences:

> The tolerance of a Norwegian beetle (Phyllodecta laticollis) to freezing temperatures varied seasonally, in association with changes in the blood concentration of glycerol, amino acids, and total dissolved solute. However, the concentration of nucleating agents in the blood did not vary seasonally.

Note that the two-sentence distillation contains considerable detail despite its brevity, implying impressive mastery of the paper's contents; it is complete, accurate, and self-sufficient.

Sample Student Summary

Minnie Leggs
Bio 101
Fall 1991

Van der Laak, S. 1982. Physiological adaptations to low temperature in freezing-tolerant Phyllodecta laticollis beetles.

Comp. Biochem. Physiol. 73A: 613-620.

Adult beetles (Phyllodecta laticollis), found in Norway, are exposed to sub-zero (°C) temperatures in the field throughout the year. In general, organisms that tolerate freezing conditions either produce extracellular nucleating agents that trigger ice formation outside the cells rather than within them or they produce biological antifreezes, such as glycerol, that lower the freezing point of the blood and tissues to below that of the environment, thereby preventing ice formation. This study was undertaken to document the tolerance of P. laticollis to below-freezing temperatures and to account for seasonal shifts in the temperature tolerance of these beetles.

Beetles were collected throughout the year and frozen to temperatures as low as –50°C; post-thaw survivorship was then determined. Determinations were also made of the concentrations of solutes in the blood (that is, blood osmotic concentration), total water content, amino acid and glycerol concentrations in the blood, presence of nucleating agents in the blood, and the temperature to which blood could be super-cooled before freezing would occur.

The temperature tolerance of P. laticollis varied from about –9°C in summer to about –42°C in winter; this shift in freezing tolerance was parallelled by a dramatic winter increase in glycerol concentration and in total blood osmotic concentration. Amino acid concentration also increased in winter, but the contribution to blood osmolarity was small compared to that of glycerol. Nucleating agents were present in the blood year-round, ensuring that ice formation will occur extracellularly rather than intracellularly, even in summer.

For beetles collected in mid-winter and early spring, blood glycerol concentrations could be artificially reduced by warming beetles to 23°C (room temperature) for about 24–150 h. When glycerol concentrations of spring and winter beetles were reduced to identical levels by warming, the spring beetles tolerated freezing better than the winter beetles; these differences in tolerance could not be explained by differences in amino acid concentrations. This result indicates that some other factors, as yet unknown, are also involved in determining the freezing tolerance of these beetles.

Analysis of Student Summary

The student has within about one page successfully distilled a seven-page technical report to its scientific essence. Note that the student used the first three sentences to introduce the topic and then summarised the purpose of the research in one sentence. The next short paragraph summarises the experimental approach taken, and the main findings of the study are then stated. No superfluous information is given; the author of this assignment provided only enough detail to make the summary comprehensible. The product glistens with understanding. Rereading the student's two-sentence encapsulation of the paper (page 214), you can see that the student was indeed ready to write the report.

THE CRITIQUE

A critique is much like a summary, except that you add your own assessment of the paper you have read. This does not mean you should set out to tear the paper to shreds; a critical review is a thoughtful summary and analysis, not an exercise in character assassination. Almost every piece of biological research has shortcomings, most of which become obvious only in hindsight. Yet every piece of research contributes some information, even when the original goals of the study are not attained. Emphasise the positive – focus on what *was* learned from the study. Although you should not linger on the study's limitations, you should point them out towards the end of your critique. Were the conclusions reached by the authors out of line with the data presented? Do the authors generalise far beyond the populations or species studied? Which questions remain unanswered? How might these questions be addressed? How might the study be improved or expanded? Demonstrate that you understand what you have read. Do not comment on whether or not you enjoyed the paper.

Sample Student Critique

Before writing the critique, the student produced this one-sentence summary of the paper.

> The egg capsules of the marine snails Nucella lamellosa and N. lima protect developing embryos against low-salinity stress, even though the solute concentration within the capsules falls to near that of the surrounding water within about 1h.

Again, note that this one-sentence summary satisfies the criterion of self-sufficiency: it can be fully understood without reference to the paper it summarises. The critique follows:

Saul Tee
Bio 101
Fall 1992

Kînehcép, N. A. 1982. Ability of some gastropod egg capsules to protect against low-salinity stress. J. Exp. Marine Biol. Ecol. 63: 195-208.

The fertilized eggs of marine snails are often enclosed in complex, leathery egg capsules with 30 or more embryos being confined within each capsule. The embryos develop for one or more weeks before leaving the capsules. The egg capsules of intertidal species potentially expose the developing embryos to thermal stress, osmotic stress, and desiccation stress. This paper describes the ability of such egg capsules to protect developing embryos from low-salinity stress, such as might be experienced at low tide during a rainstorm.

Two snail species were studied: Nucella lamellosa and N. lima. Embryos were exposed, at 10–12°C, either to full-strength seawater (control conditions) or to 10–12 per cent seawater solutions (seawater diluted with distilled water). The ability of egg capsules to protect the enclosed embryos from low salinity stress was assessed by placing intact egg capsules into the test solutions for up to 9h, returning the capsules to full-strength seawater, and comparing subsequent embryonic mortality with that shown by embryos removed from capsules and exposed to the low-salinity stress directly.

Encapsulated embryos exposed to the low salinities suffered less than 2 per cent mortality, even after low-salinity exposures of 9h duration. In contrast, embryos exposed directly to the same test conditions for as little as 5h suffered 100% mortality. All embryos survived exposure to control conditions for the full 9h, showing that removal from the capsules was not the stress killing the embryos in the other treatments. Sampling capsular fluid at various times after capsules were transferred to the diluted seawater, Kînehcép found that the concentration of solutes within capsules fell to near that of the surrounding water within about 1h after transfer.

This study clearly demonstrates the protective value of the egg capsules of two snail species faced with low-salinity stress. However, Kînehcép was unable to explain how egg capsules of these two species protect the enclosed embryos, since the capsules did not prevent decreases in the solute concentration of the capsular fluid. Although Kînehcép plotted the rate at which the solute concentration falls within the capsules (his Fig. 1), he sampled only at 0, 60, and 90 minutes after the capsules were transferred to water of reduced salinity. I think he should have sampled at frequent intervals during the first 60 min to discover how rapidly the solute concentration of the capsule fluid falls. As Kînehcép himself suggests, perhaps the embryos are less stressed if the concentration inside the capsule falls slowly.

These experiments were all performed at a single temperature even though encapsulated embryos are likely to experience fluctuation in both temperature and salinity as the tide rises and falls during the day; the study should be repeated using a range of temperatures likely to be experienced in the field. In addition, I suggest repeating these experiments using deep-water species whose egg capsules are never exposed to salinity fluctuations of the magnitude used in this study.

Analysis of Student Critique

As before, this student begins with just enough introductory information to make the point of the study clear and ends the first paragraph with a succinct statement of the researcher's goal. The methods and results are then briefly reviewed. Whereas a summary would probably end at this point, the critique continues with thought-provoking assessments by the student. Note that the student was careful to distinguish his thoughts from those of the paper's author (see pages 28–35, on plagiarism).

Concluding Thoughts

Successfully completing either type of assignment is no trivial matter. But preparing summaries and critiques is an excellent way to push yourself towards true understanding of what you read, and of the nature of scientific enquiry.

9. Preparing Papers for Formal Publication

Most of the necessary advice for writing papers has already been given in previous chapters, as the principles are the same as for essays and lab reports. This chapter mainly concerns the final details. For example, manuscripts should normally be double-spaced, unless the Instructions to Authors state otherwise, but journals sometimes require mathematical equations to be triple-spaced. Covering letters are usually single-spaced.

You are strongly recommended to read the one-page article, '*Review of Manuscripts from the Perspective of a Senior Editor*', by J.A. Bartz, in *Plant Disease*, October 1988, p831. He states that: 'Many manuscripts do not appear to have been reviewed or edited prior to submission and have numerous errors in organization, typing, spelling, composition, and grammar. These errors are likely to cause a negative bias in reviewers and editors: the bias increases the chance that the manuscript will be rejected. A few manuscripts are so poorly written that they must be revised before they can be evaluated by reviewers and editors (currently, about 10 per cent of the new submissions are in this category). Sometimes it appears that an author is attempting to use the journal's editorial board to convert a rough draft into an acceptable report.' That quotation reinforces much of the advice that has been central to this book.

Papers submitted to the editor for possible publication must conform exactly to the requirements of the specific journal you have targeted. Before beginning a manuscript, you must determine which is the most appropriate journal for your work and read carefully that journal's

Instructions for Authors, typically found at the front or back of each issue or, in some cases, at the front or back of particular issues each year. It also helps to study similar papers published in recent issues of the targeted journal. How are references cited in the text? How are they listed in the Literature Cited section? Does the journal permit subheadings in the Results section? What are the instructions about photographs? If you fail to follow the relevant instructions your paper may be returned unreviewed, and will annoy the editor and reviewers.

Before mailing your manuscript to the journal's editor, go through your work one last time and check that every reference cited in the text is listed (and correctly so) in the Literature Cited section, and that the Literature Cited section contains no references not actually mentioned in the text. You should also indicate, in the left margin, where each figure and table is first referred to, writing something like, 'Fig. 2 near here'. This helps the printer know where best to place each table, graph or photograph. If your manuscript does not make absolutely clear what are subscripts or superscripts or Greek letters, say in formulae or genetical terms, you can use notes in pencil in the margin nearby. For example, with γ_1 in the text, you could write in the margin nearby, 'Greek Gamma, subscript one'. If your printer or typewriter makes no distinction between the letter 'l' and the numeral '1', then use pencilled notes in the margin where there could be ambiguity. Take similar care with capital letter 'O' and numeral zero, '0'.

Your manuscript should be accompanied by the correct number of copies, as specified in the Instructions to Authors, along with a brief covering letter, which should read something like this:

[Start with the editor's name and address at the top left, and the date on the right at the top. Be sure to give your full postal address, telephone number, including extension, and Fax number and electronic mail address, if you have them.]

Dear Dr. Vernberg,

Please consider the enclosed manuscript entitled 'Influence of delayed metamorphosis on survival, growth, and reproduction of the marine polychaete Capitella sp. I', by J. A.

Pechenik and T. R. Cerulli, for publication in the Journal of Experimental Marine Biology and Ecology.

Thank you for your attention.

Yours sincerely,

[Signatures, then typed names of the authors.]

More detailed advice about preparing professional manuscripts is given in several of the references listed in Appendix B; see in particular the books by R. Day and W. S. Cleveland. Some journals ask for the paper to be submitted on floppy disk as well as in typescript, especially the final version after revisions at the editor's suggestion. The disk should be labelled with the journal, authors' names, paper's title, and the exact version of the word-processing program used, *e.g.* AmiPro 2.0.

10. Preparing Research Posters

A research poster is typically only about one metre by one metre. The first thing you need to do is to find out, from the meeting organiser or the meeting programme, what is the exact size and shape of the space available to you. The programme may also give details of how the poster is to be fixed to the surface to be provided, and whether the organisers provide the materials for fixing to the display boards, *e.g.* drawing pins or Velcro®-type adhesive pads.

Look very carefully at the instructions to exhibitors: there may be a limit on the length of the title, *e.g.* 30 characters (letters and spaces). There may be a specified format for setting out your name and institution. These things must be considered when you are working out how much space is left for the main part of your poster. If you are new to designing posters, have a look at some existing ones, perhaps displayed in the corridors of your department. Look at their illustrations, graphs, tables, print sizes, mounting and general design.

Normally, after the poster's title come the name(s) of the author(s), the name(s) of the institution(s) where the work has been done, an abstract, an introduction, materials and methods, results (including illustrations and data), analysis or discussion, and conclusions or summary. There is usually a dilemma: if you use a large size of print for legibility, you will get less information on the poster than if you used a smaller print size. Your inspection of other people's posters should give you a good idea of the best size of print to use – measure the print height with a ruler, if necessary.

Your poster should normally be word-processed or typed, not hand-written. With word-processing and the right printer, you can often use a large range of print sizes, say from 10 to 60 points high, where 72-point print would be an inch high. 10-point print is probably too small for posters. Use a very large size for your title, say 60-point or more. You can of course use an enlarging photocopier once or several times in succession to get still larger sizes, or a photographic technician can photograph and enlarge parts or all of your poster. Photographic prints metres long and wide can be printed and processed if the equipment is available: consult your local expert. Consider also having fairly large print for your introduction and summary, as these and the illustrations will often be all that the casual viewer looks at.

If possible, try to have one or more coloured illustrations. If only black-and-white illustrations are relevant, say from electron micrographs, make sure that they are as clear as possible. All illustrations must of course be labelled and have a scale, unless the scale is obvious, say because there is a human in the photo. Graphs must have both axes labelled; follow the instructions from Chapter 3 on legends for tables and graphs. Complicated and long tables of data, and very intricate graphs, are best avoided if possible. Just give the key data, as clearly as possible. There is usually no room for a long Materials and Methods section, nor for all the data from your thesis, nor for wordy discussions.

Look at Chapter 8 on writing summaries, as a poster presentation is more of a summary than a full paper. When writing the poster, bear in mind whether you will be with your poster when it is exhibited, as you will then be able to explain details to any questioners. Some scientific meetings have specific times when exhibitors should be with their posters to answer questions; note the times carefully. If your poster is for a general open day, rather than for a research meeting, then bear in mind the likely level of background knowledge of your poster's viewers, and simplify the science if necessary, with more definitions of terms.

Well before you are due to exhibit your poster, make a rough version to see how much space it occupies. You may find that you need to edit out large chunks to get the size correct. Make all sections as short, clear and succinct as possible. Get someone else – perhaps your supervisor if you are a research student – to look at your draft poster, to advise you on its appearance, understandability and its all-important scientific content, then revise it as necessary before producing the final version. As with other written work, the procedure is: plan ahead, think, write, revise; get one or more independent criticisms, think and revise again as often as necessary. With posters, however, you also need to be very space-conscious, and to take print-size, eye-appeal and visual readability very seriously.

If the various parts of your poster are printed on normal paper, it is usually best to mount them, singly or in groups, on thin card for ease of the final mounting on boards at the meeting, as paper can easily curl, tear and distort. Some printers will actually print directly onto card. Some people use white card for mounting; others prefer coloured card. It is always advisable to take a supply of mounting fixtures, even if the organisers are meant to supply them. Depending on how your poster is mounted, obtain a suitable large cardboard tube or a strong, large folder to protect your poster during transportation. And don't forget to collect your poster at the end of the display time!

11. Preparing Talks

Many undergraduate courses, especially in the final year, involve students giving talks to the rest of the class and to the lecturer. The topics may be fairly general or highly specific. Oral presentations of published research papers are sometimes given in class by final-year undergraduates, and by post-graduates to research groups. Research projects may also culminate in talks. Oral presentations are developed in the same way as written ones: see Chapter 2 for advice on finding appropriate literature for your topics. Talks must differ from written presentations in one important respect: a written page can be read slowly and pondered, and can be re-read as often as necessary until all points are understood; an oral report, however, gives the listener only one chance to grasp the material, unless the speaker repeats the key points. For maximum impact, talks must be very well organised, developed logically, stripped of details that divert the listener's attention from the essential points, and delivered clearly and smoothly.

GENERAL ASPECTS OF GIVING ANY TYPE OF TALK

Show your awareness of audience interests and knowledge. Tell the audience at the start how the talk relates to their interests. Think in advance about your audience's level of background knowledge and pitch your talk accordingly. With individuals varying greatly in this level, use phrases such as 'I'm sure many of you know this already, but for those who don't. . .' when you cover basic aspects for the less knowledgeable.

Make sure that you know the material you will be talking about so that you are not over-dependent on your notes. Unless you are giving direct quotations, never read your notes word-for-word, as doing so makes for a very boring talk. Glance frequently at the audience, to draw all of them into your talk; do not just look at your friends. Eye-contact with the audience is a key factor in giving lively talks, so do not spend long periods talking to the board or to your notes. Project your voice as if speaking directly to someone in the back row.

During early preparation, think about logical sequences. Which facts must come early in the talk to provide the background for later ones? Tell the audience at the start what the main purpose of the talk is, and how you are going to deal with the subject, perhaps using an overhead projector (OHP) transparency to show a plan of the talk's subdivisions. During the talk, write subheadings on the board or on OHP transparencies to let the audience know which section you have reached. Using a series of numbered points helps make the organisation clear.

If your talk seems to consist of a flat series of statements, introduce variety by asking and answering questions. If you can introduce some relevant specimen or small demonstration, that adds to the sense of occasion. Beware of having long sequences of slides in a darkened room, as that can induce sleepiness in the audience. If the audience is likely to want to take notes, allow extra time for that, with pauses at suitable moments, instead of your dashing from one OHP transparency or slide to the next. Some light is necessary for note-taking. Don't be depressed by rows of bowed heads if the audience is seriously taking notes. Do use your voice, gestures and phrases such as 'This is of crucial importance. . .' to distinguish between really important matter and less important matter.

Your manner of speech should be friendly and helpful, not dictatorial, aggressive or defensive. If necessary, use a microphone, as shouting makes the voice sound aggressive. Vary the tone, volume and speed of your speech, *e.g.* slowing down to emphasise the most important points. Be neither static nor over-mobile when speaking. Use one hand to gesture, point or write, if you hold your notes in the other hand. Do not gable desperately if you run short of time: cut out details and summarise the rest of the material. You are very strongly advised to have a rehearsal of your talk well in advance, perhaps with a friend as an audience to give you advice, and to help you get the timing right. If you over-run your time at the private rehearsal, cut out some details rather than speed up the delivery. Before preparing all your materials, such as OHP transparencies or slides, experiment privately to find out what size of lettering is needed for full visibility from the back of the room. No one likes to see overcrowded slides full of unreadably-small print.

Do not worry if you feel nervous before you start your talk, with sweaty palms or a need to visit the lavatory. These signs show that your body and mind are fully prepared for action! It is better to be nervous and make a real effort to give the talk well, than to be over-relaxed and make an inadequate effort to get the subject across to the audience. Just concentrate on putting your message as clearly as possible. Do not worry about what the audience might be thinking about you: they will appreciate your efforts on their behalf if you are well-prepared and sincere.

Know what you're going to say and how you're going to say it. Hesitation, vagueness and searching for words will all suggest a lack of understanding and, in addition, will lose the attention of your audience. Write out your talk in legible note-form, and practise it until you can produce a smooth delivery. Don't rush. Write on the blackboard or overhead transparencies during your presentation, for example, when labelling the axes of graphs. This helps punctuate your statements and also gives the listener time to digest what you are showing as well as time to take notes. For the same reason, you should label curves as you draw and talk about them. It is often a mistake to put completed illustrations on the blackboard in advance; the listener may be deprived of the opportunity to absorb what is being presented.

Make the blackboard work for you by drawing the listeners' attention to specific aspects of the graphs and tables that represent the point you wish to make. Don't simply say, 'This is shown in the graph on the board.' Rather, say 'For example, all the animals fed on diets *A* and *B* grew at comparable rates, but . . . ,' and be sure to point to the data as you speak.

If using an overhead projector, point to the screen when you wish to highlight a detail, not to the transparency itself. Transparency projectors magnify, and will transform the natural, barely noticeable nervous tremor of your hands into a highly distracting display of stage fright. If possible, try out the equipment in advance, finding out how to switch projectors on and off, how the focus controls operate, how to dim and raise the room lights, how to raise or lower the screen, and any other necessary technical details. If making OHP transparencies by photocopying, make sure that your OHP sheets are safe in your photocopier, as some acetates just melt messily inside the photocopier.

Write unfamiliar terms on the blackboard or overhead transparency and avoid acronyms whenever possible; there is no justification for referring to 'NCAM's' instead of to 'neural cell adhesion molecules' when the term is used only once or twice in the talk. Your goal is to communicate, not to impress or confuse.Try to sound interested in what you are saying, no matter how many times you have practised your talk. If you seem bored, you will certainly bore the audience.

At the end, summarise the main points of the talk, briefly but clearly. Don't stop abruptly. Warn your audience when you are nearing the end of your talk by saying something like, 'I would like to make one final point', or 'Before I end, I wish to emphasise that . . . .'

End your talk gracefully. A self-conscious giggle or a 'Well, I guess that's it' isn't the best way to finish. Try something like, 'Thank you. I would be happy to answer any questions.' Do not feel compelled to answer questions that you don't understand. Politely ask for clarification if you are unsure what has been asked. Paraphrase each question before answering it so as not to lose the rest of the audience (and to give yourself time to think). 'The question is, does the technique used to isolate the DNA interfere with. . . .' Then address your answer to the entire audience, not just the questioner.

TALKING ABOUT PUBLISHED RESEARCH PAPERS

A talk can be effective only if you fully understand your topic. It is wise to skim the paper that you are presenting once or twice for general orientation. Consult appropriate textbooks for background information if necessary. Pay particular attention to the Materials and Methods section and to the tables, graphs and photographs in the Results section. When you can summarise the essence of the paper in one or two sentences, you are ready to prepare your talk. Try to capture the essence of the research – why it was undertaken, how it was undertaken, and what was learned – and to communicate that essence clearly, convincingly and succinctly.

Do not simply paraphrase the paper or papers that you are presenting if you wish to keep your audience awake: you must reorganise the information. Begin your talk by providing background information, drawing from the Introduction and Discussion sections of the paper and from outside sources if necessary, so that the listener can appreciate why the study was undertaken. End your introduction with a concise statement of the specific question or questions addressed.

Be selective; delete extraneous details and focus your talk on the results. Much of what is appropriate in a research paper is not necessarily appropriate for a talk about it. Some of the details – particularly of methods – must be pruned. Include only those details needed to understand what comes later.

Draw conclusions as you present each component of the study, so that you lead in logical fashion to the next point addressed. For example: 'The oyster larvae grew 20 μg/day when fed diet *A*, 25 μg/day when fed diet *B*, and 65 μg/day when fed a combination of diets *A* and *B*. This suggests that important nutrients missing in each individual diet were provided when the diets were used in combination. To determine what these missing nutrients might be. . . .' Lead your audience from point to point.

Plan to use the blackboard or overhead transparencies. A simple summary table or two is helpful when numbers are being discussed; numbers floating around in the air are very difficult for listeners to keep track of. A diagram of experimental protocol can help the listener follow the plan of a study. Data can often be summarised in a few graphs, even when those data were presented in the original paper as complicated tables. Keep the graphs simple and label both axes. You need not reproduce graphs exactly as given in the paper and you need not display every entry from a particular table. Focus on showing the trends in the data.

Summarise the major findings of the research at the end of your talk, driving the points home one by one. You may wish to end with a brief discussion of the way the study could be improved or expanded, but don't set out to discredit the authors. End on a positive note, reinforcing what you want your audience to remember. Be prepared for questions about methodology. Listeners often ask about interpretations of the data; to answer these questions, you must be thoroughly familiar with the way the study was conducted.

TALKING ABOUT ORIGINAL RESEARCH

Follow the format described above for preparing and presenting your work. Again, begin by presenting the background information that listeners need in order to understand why you addressed that particular question and clearly state the specific question being considered. Focus on the results of previous studies when presenting background information and on your own results when giving the rest of the talk. Draw your conclusions point by point as you discuss each facet of the study, showing how each observation or experiment led to the next aspect of the work. End the talk by summarising your major findings with their potential significance, and give some suggestions for future work.

TALKING ABOUT PROPOSED RESEARCH

This is similar to presenting a research paper except that you have more literature to review. Highlight a few key papers that show particularly clearly why the question you wish to address is a worthwhile and logical one, and again focus on the results of the studies you discuss. Then state the specific question you plan to address in your own work, being sure this follows logically from the work you have just summarised. Finally, describe the approach you will take, what you will do, and make clear what each step of the study is designed to accomplish. Conclude by briefly summarising how the proposed work addresses the question under consideration.

THE LISTENER'S RESPONSIBILITY

Few things in life are more disappointing than putting your heart and soul into preparing and delivering a talk to an audience that appears to show no interest. When you are a member of that audience, you bear a responsibility to listen closely, and to *show* the speaker that you listened closely. Try to formulate at least one question by the end of the talk, about something you didn't understand, something you thought was particularly interesting, or something unusual you saw in the data. Even if the speaker can't answer your question, he or she will at least detect some interest in the talk and feel flattered that you cared enough to have paid so much attention.

12. Writing Letters of Application

An application for a job, or for admission to a post-graduate or professional programme, will generally include a Curriculum Vitae (CV) and an accompanying covering letter (both of which you write), and several letters of recommendation (which you do not see) from your referees. Your CV summarises your educational background, any work experience or research experience, goals and general interests. The covering letter identifies the position for which you are applying and draws the reader's attention to those aspects of your CV that make you a particularly worthy candidate. The recommendations give an honest assessment of your strengths and weaknesses (we all have some of each) and offer the reader an image of you as a person and as a potential employee or post-graduate student.

BEFORE YOU START

Try to put yourself in the position of the people who will read your application. What will they be looking for? They will probably have three main questions in mind:

1. Is the applicant qualified for this particular position?
2. Is the applicant really interested in our programme or company?
3. Will the applicant fit in here?

Your application must address all three issues. There may be many applicants for the position and it is up to you to make the best of your record and skills, to convince people that you are a suitable candidate for interview and further investigation.

Make full use of any help offered by your departmental, college or university careers officer or service. They, your university library or your local public library will probably have useful books and background information, such as guides to what research and what post-graduate teaching are done at different institutes, and guides to commercial and industrial firms. Examples are *How to Write a Curriculum Vitae* (1993, University of London Careers Service, London), *Postgraduate Studies in Britain* (Newpoint Publishing Co. Ltd., East Grinstead, published annually) and *Kompass Register* [of firms], Kompass, East Grinstead, updated annually.

Do learn as much as you can about the department, institute or firm you are applying to, before you write either your CV or the covering letter, so that you can suit what you write to that particular place. Ask a firm for a detailed job description if one is not provided. Many job descriptions are given in ROGET (the *Register of Graduate Employment and Training*, C.S.U. Publications, Manchester, updated regularly). With a detailed job description, you can see whether the job would suit your abilities and training, and you can emphasise all those points of your experience most related to the required skills.

PREPARING THE CURRICULUM VITAE

The people who read your application will probably only examine your CV for a minute or two if they have a large pile of applications. An effective CV is therefore well-organised, neat and as brief as possible, not exceeding two pages in length. There is no standard format for a CV; the model given in Figure 32 (page 237) should be modified in any way that emphasises your particular strengths.

All CVs must contain the following components if relevant:

1. **Personal details:**
   (i) Full name.

(ii) Address, telephone number and dialling code, with extension number if there is one; give home and term addresses if different, with the dates at which you will be at each address.
(iii) Date of birth, sex (unless obvious from the forenames) and marital status.
(iv) Nationality.

2. **Education details:** these can be listed in the order given, or in reverse order if you wish to emphasise your latest qualifications. Mention major prizes or awards, *e.g.* for the best degree in physiology.
(i) Schools or other institutions attended between primary school and higher education, with dates and main exam results, *e.g.* 8 subjects GCSE at grades C or better, including Biology (A), Maths (B) and Chemistry (B) – the subjects you name should be those relevant to the job, and give the highest grades first. Name all the A-levels or equivalents and give grades.
(ii) Higher education institutes, with dates; degree or equivalent title, subject and class (actual or expected: state which); final year project or dissertation title. State, in your CV or the covering letter, the titles of any courses taken which are particularly relevant to the job applied for. Include details of any industrial placements or work abroad which were part of the degree.
(iii) Post-graduate education: institutes or firms, with dates; titles of higher degrees or diplomas, and whether by taught courses, research or both; any industrial or extra-mural work as part of the qualification; thesis title.

3. **Work experience**, if any: include voluntary work as well as paid jobs. A list can be in forward or reverse time sequence, or grouped by type of work: name of employer, dates, job title and duties; skills gained, *e.g.* major techniques learned, dealing with different types of people, financial or management responsibility, experience of industry, commerce, retail trade, government, academia, etc.

4. **Other interests:** briefly list these, especially if they show that: you have broad interests, a lively mind, a well-rounded personality, a bal-

anced life, and have mixed successfully with a wide range of different types of people, or if skills acquired in following these interests relate to the job applied for. For example, you might have run a voluntary society, showing abilities to administer, get on well with people, show financial control, fund-raising abilities, etc. Positions of responsibility (chairman, secretary, treasurer, fund-raiser) are worth mentioning. Also mention cultural activities such as music or drama, sports, and other suitable leisure activities.

5. **Specific skills:**
   (i) Major scientific techniques of which you have detailed experience.
   (ii) Amount of fluency, written or oral, in particular foreign languages.
   (iii) Keyboard, typing, word-processing skills.
   (iv) Computing skills: list the programming languages and major applications packages you can use.
   (v) Driving experience – length of time; mention it if your licence is 'clean', and whether you have experience of, and any necessary licence for, non-car vehicles, *e.g.* Heavy Goods Vehicles, tractors; pilot's licence; sailing experience. If applying for research or a job in marine biology, then you would emphasise in your CV and covering letter any aquatic/marine experience, and sailing or diving skills.

6. **Referees:** either in the CV or in the covering letter, list the names of two or three available people who have agreed to write references for you. They should include such people as your personal tutor and your project supervisor, or a boss or other senior person who knows you well. State the capacity in which they know you, *e.g.* project supervisor, and perhaps their position, *e.g.* Professor of Bacterial Physiology, Consultant Haematologist, Director of Undergraduate Studies, Head of Accounts Section.

In your CV, avoid drawing attention to any potential weaknesses; if, for example, you lack teaching experience, do not write 'Teaching experience: none.' Use the CV exclusively to demonstrate your strengths. You may include a one- or two-sentence statement of your immediate and long-range goals, if known. You should alter your CV for each application completed, to focus on the different strengths required by different jobs or courses.

FIGURE 32.

Sample Curriculum Vitae.

CURRICULUM VITAE
DAVID JAMES ROBINSON, BSc(Hons)
40 Medcroft Gardens,
East Sheen,
London SW14 6PL
Tel. 081-8768920.
(Same address, home and term-time.)

PERSONAL DETAILS	Date of birth	4 April 1971.
	Nationality	British.
	Marital status	Single.

EDUCATION

1982 to 1989	Tiffin Boys' School, Kingston-upon-Thames. 8 GCSE's at grade C or better, including Maths (A), Chemistry (B), Physics (B), Biology (B), English (C) and French (C). A-levels in Biology (A), Chemistry (B), Maths (C).
1989-1992	Imperial College, London University, Biology Department. BSc, 2(i) degree in Microbiology. [*] Final year project on 'The effect of DNA supercoiling on the transcription of bacterial respiratory proteins'.
CAREER OBJECTIVES	I would like a career in medical microbiology, as that could be intellectually interesting and challenging, as well as of benefit to disease-sufferers. Ideally, I would like experience of the subject in academic, hospital and pharmaceutical industry environments.

POST-GRADUATE WORK, 1992-93	I am now taking a one-year MSc Course in Medical Microbiology at St. Mary's Hospital Medical School, Paddington, and will probably do a mycological dermatology research project this summer.
WORK EXPERIENCE	Voluntary work with patients at Queen Mary's Hospital, Roehampton, 1987-88. Christmas vacation, 1990, 2 weeks at Mortlake postal sorting office, meeting an interesting cross-section of the population. Summer, 1991, 3 months work as student scientific officer at the Imperial Cancer Research Fund, Lincoln's Inn Field, London, on the Human Genome Analysis Project, getting practical experience of techniques used to make and screen a clone bank, with YAC and COSMID vectors, and other molecular methods.
OTHER INTERESTS	Rugby, for school 1st team and Imperial College 2nd and 3rd teams. Reading modern fiction. Writing science fiction. Chamber music. Social aspects of medicine, especially overseas.
OTHER SKILLS	Word-processing (Word Perfect and AmiPro), Graphics (Lotus Freelance), Programming (Fortran and Pascal). First aid (St John's Ambulance Certificate). Good oral and written French. Clean driving licence since 1990. *[Details of the courses taken which are most relevant to the application would be included here.] [The names of referees could be given here at the end of the CV, or in the covering letter.]

PREPARING THE COVERING LETTER

The covering letter plays a large role in applications and is usually the first part of your application read by an admissions committee or prospective employer. A well-crafted letter can help to counteract a mediocre academic record. Keep revising this letter until you feel that it works well on your behalf. Have some friends and perhaps your personal tutor read and comment on it, then revise it again. Be sure to type the final copy very neatly. Many employers bewail the frequent poor standards of presentation, spelling and grammar in applications they receive. The covering letter should be about one or, at most, two typed pages.

Do not simply write,

> Dear Mrs. Harris,
>
> I am applying for the position advertised in The New Scientist. My CV is enclosed. Thank you for your consideration.
>
> Yours sincerely,
>
> John Carpenter

Although the letter ends well, its beginning is vague and its mid-section does little to further the applicant's cause. In the covering letter:

1. At the top left, type the name and address of the person to whom you are applying, and the date at the top right.
2. State where and when the job advertisement was published, and give any reference numbers specified. Identify the specific position or course for which you are applying.
3. Draw the reader's attention to those elements of your CV that you feel make you a particularly qualified candidate.
4. Indicate that you understand what the position entails, that you have most of the necessary skills, and that you are eager and able to learn any additional necessary skills.
5. Convince the reader you are a mature, responsible person, not flippant or casual.
6. Convey a genuine sense of enthusiasm and motivation.

Before you write the letter, ask yourself some difficult questions:

- Why do I want this particular job or to enter this particular post-graduate course?
- What skills would be most useful in such a job or course?
- Which of these skills do I have?
- What evidence of these skills can I present?

Your answers to these questions will provide the basis for your covering letter. Tailor each letter to the particular position or course for which it is being prepared. Try to find some special reason for applying; if possible, your application should reflect deliberate choice and a clear sense of purpose. If, for example, you have read papers written by a staff member at the institution to which you are applying and have become interested in that person's research, weave this information into your covering letter, and be prepared to discuss that work at an interview. On the other hand, if your major reason for wanting a particular job or to attend a particular post-graduate programme is the geographic location of the company or university, be careful not to state this as your sole reason for applying; try to put yourself on the receiving end of the covering letter and consider how your statements might be interpreted.

Back up statements with supporting details. Avoid simply saying that you have considerable research experience. Instead, briefly explain what your research experience has been. State the facts and let the reader draw the proper inferences.

A specimen CV for a Mr Robinson was given above. His covering letter would be different for applications for PhD places, industrial jobs and hospital posts. For all of these, he would stress any undergraduate courses relevant to the proposed area of work, listing them in his CV or the covering letter. So for a medical post, he might mention a general background in biology and biochemistry, with supporting maths, statistics and physics, and specifically mention second year Molecular and Cell Biology, Immunology, Mycology, Bacterial Physiology, Virology, and Computing, and third year Medical Microbiology (taught at St Mary's Hospital Medical School), Applied Genetics, and Epidemiology. His voluntary work in a hospital, his scientific vacation job, and the relevance of his undergraduate project and his post-graduate course would all be briefly referred to in the covering letter, in case the reader overlooked such helpful material in the CV.

For an academic PhD, his academic record would be stressed, plus any details relevant to skills or knowledge needed for the proposed PhD, including his research experience so far. For an industrial job, relevant skills and experience would again be mentioned, plus any managing experience, for example, running a society, and any business experience, even just taking part in business-games at school, or industrial visits.

OBTAINING EFFECTIVE LETTERS OF RECOMMENDATION

Letters of recommendation can be extremely important in determining the fate of your application. Although you do not write these letters and rarely get the opportunity to read them, you can take steps to increase their effectiveness.

Getting an *A* in a course does not guarantee a strong letter of recommendation from the lecturer. The most useful letters are those commenting on characteristics such as: intellectual ability, originality, lab skills, communication skills (written and oral), motivation, ability to use time efficiently, ability to learn or devise new techniques, curiosity, maturity, leadership potential, ability to work independently, and ability to work with others. We are sometimes also asked about your health, attendance, punctuality, responsiveness, honesty and even your mental stability! We cannot comment on all these attributes unless you became well-known to us, not just one name in a huge list of students. Make an appointment to talk with people like your personal tutor about your interests and plans. Academic staff are usually happy for the opportunity to get to know students better.

Refer back to page 236 for what kinds of person to choose as referees. Ask each of them if he or she would be able to support your application by writing letters of recommendation. Give each person the opportunity to decline your invitation. If the people you ask agree to write on your behalf, make their task easier by giving them a copy of your CV, any transcript or academic record of your performances on each course, your letter of application, and, if appropriate, a copy of the job advertisement. Be certain to indicate clearly the application deadline and the address to which the recommendation should be sent – a stamped, addressed envelope from you can save your referee's time.

Don't lose good will by requesting letters at the last minute. Give your referees at least two weeks to work on these letters, especially on a first letter for you. 'It has to be in by this Friday' will annoy your prospective advocate and may not allow the recommender the time needed to prepare a good letter – even if he or she is still in a cooperative mood. Moreover, last-minute requests don't speak favourably about your planning and organising abilities, and imply a lack of respect for your referee.

13. Practice at Spotting and Correcting Errors

Try to spot *all* the errors in these genuine quotations from non-dyslexic, UK-educated writers. Decide what are the errors of science and of English, and how best to correct them. You should be able to do this even if you do not know the relevant biology. Check your answers with Appendix C, pages 254–5.

1. One bacteria and two fungus were plated on three mediums as seperate colonies.
2. The price that a breeder might get for his cow would depend on how many prizes he had got and not on his siring ability.
3. There is still plenty of sugar left behind mainly fructose the sweeter of the 2.
4. The results reveal a clear asymmetry, and the significance of the results are discussed.
5. Produced mainly in France at figures of .3,577;200 pounds.
6. It is regulated by the formation of less toxic waste products than their acquatic ancesters.
7. This book is based on lectures and clinical demonstrations of venereal diseases which the author has been giving to the undergraduate and postgraduate students for many years.
8. [From a dictionary] Free electron – an electron in a crystal which is free to move under the influence of an electrical field.
9. The above affects of enviroment all consider the external enviroment.

10. I am happy to comment on the manuscript but would rather the authors did not know the sauce of the comments.
11. Replacement gene therapy by the use of cell targeted retro-viruses bearing the wild type gene may emerge from the world of science fiction into the world of fact to elevate this most interesting of deseases.
12. His first law states that he thought characters where past on from generation to generation by particulate factors.
13. When lactose is present, the amount of enzyme jumps up to maybe 50,000.
14. The electron microscope has to be in a very stable room to avoid vibration which is acheived by concrete bases.
15. The male possesses sex combs on its front legs. These are not present in the female.
16. The anther was mounted on a slide in iodine to test for starchy and non-starchy.
17. The method used depends on the type of pollination the crop undergoes i.e. wheather self-pollination, cross-pollination or vegetative reproduction.
18. There are a few cases of men with XYY chromosomes and these tend to be tall, powerful and aggressive.
19. Short plants will only be produced in the absence of tall genes.
20. In moths, half the gametes produced by the female are male gametes, half are female gametes.
21. The solanifolia variety has low fertility, due to a deformed pistol and so may be selected against through the prescence of a highly fertile variety, able to establish itself quicker.
22. Incubating at 25° stimulates room temperature and 7 colonies where propagated.
23. Considering the metabolism of phenylalanine and tyrosine several well characterised genetic defects arise in this way.
24. Plate (c) shows that tryptophan reproduces normally at lower temperatures.
25. Using the sterile velvit tecnique colonys are then transfered to different agar plates.
26. This is the herb which is used to cure the piles and anus irritation which contain pericarp like seeds.

27. The fate of unused embryos can be freezed for a future attempt.
28. The sex, birth date, birth weight and birth rank is recorded.
29. It is desireable that human marriages are prohibitive between close relatives.
30. The fact of having sexes infers the process of sexual reproduction.
31. By maintaining only low levels of these enzymes; until a suitable resource is encountered; the bacterium reduces the energy involved in protein synthesis.
32. None of the mutations were very delaterious.
33. Even number ploidys generally produce fertile plants.
34. Albinism effects 1 in 20,000 live births.
35. These females need counselling against pregnancy and sterilisation or abortion may be practised.
36. Into the tube that contains the Conidia, pour a small amount of dilute water, with a innoculating loop.
37. Were cross-feeding occured, growth occured as well.
38. The cells of a pentaploid newt are roughly five times the size of a haploid newt although the overall size of the organism is unchanged.
39. The amount of cytoplasm in these cells increases to preserve the ratio of the size of the nucleus in comparison to the size of the cytoplasm. (A statement which matches the size of the pentoploid newts.)
40. This carries on till a stop codon. Were The system dissasembles. And a protein is got.

You should also look at Figure 2, page 30, line 3 of the main text, first word, 'has', and find this verb's subject.

APPENDIX A

MEANS, VARIANCES, STANDARD DEVIATIONS AND STANDARD ERRORS

Suppose you have two samples of three rats each. The rat tail lengths in samples *A* and *B* are:

A = 7.0, 7.0, 7.0 cm
B = 3.6, 14.1, 3.3 cm

Both samples have the same mean value (7.0 cm), but *A* is much less variable than *B*. Simply listing the mean value, then, omits an important component of the story contained in your data.

The *variance* (σ^2) about the mean gives an indication of how variable your data are from one observation to the next. While the mean here has a value in cm, the variance will have the units cm^2. If you have access to a statistical calculator, just push the right buttons. If not, make your calculations using this formula:

$$\sigma^2 = \frac{\sum_{i=1}^{N} (X_i - \overline{X})^2}{N - 1}$$

N is the number of observations made, X_i is the value of the ith observation and $\overline{X}$ is the mean value of all the observations made in a sample.

Σ is the symbol for summation. In this case, you are to sum the squared differences of each individual measurement from the mean of all the measurements. As an example, suppose you have the following data points:

5 cm
4
4
6
5

N = 5

$$\overline{X} = \frac{\sum_{i=1}^{N}}{N} = \frac{24}{5} = 4.8 \text{ cm}$$

$$\sigma^2 = \frac{(5 - 4.8)^2 + (4 - 4.8)^2 + (4 - 4.8)^2 + (6 - 4.8)^2 + (5 - 4.8)^2}{4}$$

$$= 0.7 \text{ cm}^2$$

All you are doing is seeing how far each observation is from the mean value obtained and adding all these variations together. The squaring eliminates minus signs, so that you have only positive numbers to work with. Clearly, 100 measurements should give you a more accurate estimation of the true mean tail length than only ten measurements; 1000 measurements would be better still. We thus divide the sum of the individual variations by a factor (N–1) related to the number of observations made. Increasing the sample size will reduce the extent of experimental uncertainty. Variance, then, is a measure of the amount of confidence we can have in our measurements. The smallest possible variance is zero (all samples were identical); there is no limit to the potential size of the variance.

To calculate the standard deviation (SD), simply take the square root of the variance. To calculate the standard error of the mean, simply divide the standard deviation by the square root of N. The standard error of the sample mean, $s_{\overline{x}}$, is an estimate of the standard deviation of the means of many samples which might be taken from a population; it is a measure of the closeness with which the sample mean, $\overline{X}$, represents the population mean.

APPENDIX B

SUGGESTED REFERENCES FOR FURTHER READING

Guides to Books

Whitaker's Books in Print. Whitaker (J) & Sons Ltd, London. Updated yearly; also available on microfiche. Specialises in UK books.
Books in Print. R.R. Bowker/Reed International, NJ. Updated regularly; also available on microfiche. Specialises in USA books.

About Writing

Barnet, S. and M. Stubbs. 1990. *Practical Guide to Writing,* 6th ed. Scott, Foresman and Co., Glenview, IL.
Hall, D. and S. Birkerts. 1991. *Writing Well,* 7th ed. HarperCollins Publishers, Inc., New York.
Kahn, J.E., Editor. 1991. *How to Write and Speak Better.* The Reader's Digest Association Limited, London.
Strunk, W., Jr. and E. B. White. 1979. *The Elements of Style,* 3d ed. The Macmillan Co., New York.

Scientific Writing

Day, R. A. 1988. *How to Write and Publish a Scientific Paper,* 3d ed. Oryx Press, New York.
Gopen, G. D. and J. A. Swan. 1990. '*The Science of Science Writing.*' *American Scientist* 78: 550-558.
King, L. S. 1978. *Why Not Say It Clearly? A Guide to Scientific Writing.* Little, Brown and Co., Boston.
Lindsay, D. 1984. *Scientific Writing.* Longman Cheshire, Melbourne.
O'Connor, M. 1991. *Writing Successfully in Science.* Chapman & Hall, London.

Shortland, M. and J. Gregory. 1991. *Communicating Science.* Longman Scientific & Technical, Harlow.

Wilkinson, A. M. 1991. *The Scientist's Handbook for Writing Papers and Dissertations.* Prentice-Hall, Englewood Cliffs, NJ.

Zinsser, W. 1991. *On Writing Well: An Informal Guide to Writing Non-fiction,* 4th ed. HarperCollins Publishers, Inc., New York.

Technical Guides for Biology Writers

Biological Nomenclature: Recommendations on Terms, Units and Symbols. 1989. Institute of Biology, London.

CBE Style Manual Committee. 1978. *Council of Biology Editors Style Manual: A Guide for Authors, Editors, and Publishers in the Biological Sciences,* 4th ed. Council of Biology Editors, Washington, D.C.

Jerrard, H.G. and B. McNeil. 1992. *A Dictionary of Scientific Units.* Chapman & Hall, London.

Aspects of English

Crystal, D. 1988. *Rediscovering Grammar.* Longman, Harlow.

Davies, C. 1981. *Brush up your Grammar.* Solo Publishing, Christchurch.

*Davis, J. 1989. *Handling Spelling.* S. Thornes, Cheltenham.

Dummett, M. 1993. *Grammar and Style.* Gerald Duckworth & Co. Ltd., London.

*Gee, R. and C. Watson. 1990. *The Usborne Book of Better English.* Usborne Publishing, London.

M.H. Manser, Editor. 1990. *Bloomsbury Good Word Guide,* 2nd ed. Bloomsbury Publishing Ltd., London.

Temple, M. 1989. *A Pocket Guide to Written English.* John Murray, London.

Thomson, A.J. and Martinet, A.V. 1986. *English Grammar.* OUP, Oxford.

Waldhorn, A. and A. Zeiger. 1991. *English.* Made Simple Books. Butterworth-Heinemann, Oxford.

*Mainly for children, but still very useful.

English Standards of Biology Students

Lamb, B.C. 1992. *'Spelling standards amongst undergraduates.'* In: *Conference '92; Reading, Spelling and Sex Education.* N. Seaton, Editor, 14: 6-14. Campaign for Real Education, York.

Lamb, B.C. 1992. *A National Survey of UK Undergraduates' Standards of English.* The Queen's English Society, London.

Richardson, J. and R. Lock. 1993. *'The readability of selected A-level biology examination papers.' Journal of Biological Education* 27: 205-212.

Constructing Effective Graphs

Cleveland, W. S. 1985. *The Elements of Graphing Data.* Wadsworth Advanced Books and Software, Monterey, CA.

Statistics

Bland, M. 1987. *An introduction to medical statistics.* Oxford University Press, Oxford. (Corrected reprint 1993.)

Sokal, R.R. and F.J. Rohlf. 1981. *Biometry: the Principles and Practice of Statistics in Biological Research.* 2nd. ed. Freeman, Oxford.

Fowler, J. and L. Cohen. 1990. *Practical Statistics for Field Biologists.* Open University Press, Milton Keynes.

APPENDIX C

EXERCISE ANSWERS AND REVISED SAMPLE SENTENCES

Words corrected for spelling are underlined.

From page 156

1. Comma should be semicolon or colon (run-on sentences).
2. Second comma of a pair needed, after 'bacteria'.
3. Punctuation changes are not enough for good scientific sense here, so some words have been changed too, but the whole sentence needs rewriting: If non-nutrient-requiring genes [genes giving no added nutrient-requirement] are transduced into special-nutrient-requiring bacteria, co-transduction can be detected by tests for the nutrients required.
4. Semicolon stops the sense and should be a comma.
5. . . . cell-to-cell contact; the DNA, which will be integrated, is in the media.
6. The apostrophe should be in 'mother's', not 'hormones'.
7. Comma needed after 'bias'. mistranslation.
8. . . . without aphids' parthenogenetic reproduction, insecticide-resistance would spread. . . .
9. No apostrophes wanted; comma after 'occurs'.

From page 160(3)–161

i, definition; a, radiation; a, irritation; e, environmental; a, degrade.

From page 161(7(i))–162

clonable; peaceable; forgiving; manageable; managing; stony; coding.

From page 162 (7(**iv**))

disbudded; occurred; differed; transmitted; targeted; ordered; preferred; propelled; committed; labelled.

From page 162 (**V,d**)

necessarily; laziness; happiness; displayed; friendliest; burying.

From pages 203–204

1. To perform this experiment there had to be a low tide. We conducted the study at Blissful Beach on September 23, 1991, at 2:30 PM.
2. In *Chlamydomonas reinhardi*, a single-celled green algae, there are two mating types, + and –. The + and – cells mate with each other when starved of nitrogen and form a zygote.
3. Protruding from this carapace is the head, bearing a large pair of second antennae.
4. The order in which we think of things to write down is rarely the order we use when we explain what we did to a reader.
5. The purpose of Professor Wilson's book is the examination of questions of evolutionary significance.
6. Swimming in fish has been carefully studied in only a few species.
7. One example of this capcity is observed in the phenomenon of encystment exhibited by many fresh water and parasitic species.
8. An estuary is a body of water nearly surrounded by land whose salinity is influenced by freshwater drainage.
9. In textbooks and many lectures, you are being presented with facts and interpretations.
10. The human genome contains at least 50,000 genes, however there is enough DNA in the genome to form nearly 2×10^6 genes.
11. These experiments were conducted to test whether the condition of the biological films on the substratum surface triggered settlement of the larvae.
12. Various species of sea anemones live throughout the world.
13. This data clearly demonstrates that growth rates vary with temperature.
14. Hibernating mammals mate early in the spring so that their offspring can reach adulthood before the beginning of the next winter.
15. This study pertains to the investigation of the effect of this pesticide on the orientation behavior of honey bees.
16. The results reported here have led the author to the conclusion that thirsty flies will show a positive response to all solutions, regardless of sugar concentration (see Figure 2).
17. Numbers are difficult for listeners to keep track of when they are floating around in the air.

18. Measurements of salamander respiration ~~by the salamanders~~ typically took one-half hour each.
19. ~~The~~ results suggest that ~~some local enhacement of~~ pathogen specific antibod~~y~~ies are produc~~tion~~ed at the infection site ~~exists.~~ and thus are enhanced locally.
20. ~~Usually it has been found that higher temperatures (30°C) have resulted in the production of females, while lower temperatures 22–27°C) have resulted in the production of males. (e.g., Bull, 1980; Mrosousky, 1982)~~. The turtles are typically born females when embryos are incubated at 30°C, and male when incubated at lower temperatures (22–27°C) (e.g., Bull 1980; Mrosousky, 1982).

From pages 243–245

1. One bacterium and two fungi . . . media . . . separate.
2. ‘he’ and ‘his siring’ were presumably meant to refer to ‘cow’, not ‘farmer’, but the faults of ambiguity are compounded by an ignorance of the fact that ‘cows’ are female!
3. behind, mainly fructose, the sweeter of the two.
4. significance . . . is. . .
5. Not a complete sentence and the figure has wrong punctuation within it.
6. aquatic ancestors; faulty comparison – than those of their. . . .
7. Ambiguity. Did he give diseases to the students?
8. Ambiguity; it is not the crystal which is free to move.
9. environment (twice); effects, and effects can not consider.
10. source.
11. Hyphens needed in ‘cell-targeted’ and ‘wild-type’; commas needed after ‘therapy’ and ‘wild-type gene’; ‘elevate’ should be ‘alleviate’; diseases.
12. were passed. His law does not state: ‘I think. . .’
13. Much too vague and unscientific. . . .increases to about 50,000 molecules per cell.
14. achieved. Ambiguity as to whether vibration is achieved or cured by these bases.
15. Ambiguity of ‘These’.
16. Iodine is a solid – a solution (in what?) should be mentioned. The anther, not the slide, was in the solution, and the final adjectives need a noun (grains).

17. Comma after 'undergoes'. whether. Vegetative reproduction is not a type of pollination.
18. Ambiguity of 'these'.
19. Genes determining tallness.
20. There is confusion between female gametes and female-producing gametes, etc.
21. pistil, presence. More quickly.
22. simulates, were. seven.
23. 'Considering . . . tyrosine' is a 'hanging phrase' – see page 174–5. Comma after 'tyrosine'. well-characterised.
24. The amino acid tryptophan cannot reproduce. Tryptophan-requiring strain?
25. velvet, technique, colonies, transferred; comma after 'technique'.
26. pericarp-like seeds were from the herb, not contained in the irritation.
27. frozen. One cannot freeze a fate.
28. are recorded.
29. desirable; prohibited.
30. Facts cannot make inferences, but they can imply.
31. Semicolons block the sense; commas are needed.
32. was, deleterious.
33. Even-number ploidies.
34. affects. Note the complete change in meaning.
35. Comma needed after 'pregnancy' to remove ambiguity as to whether sterilisation is practised or counselled against.
36. conidia, an, inoculating; what is dilute water, and how do you pour it with a loop?
37. Where, occurred (twice).
38. Is size here diameter, surface area or volume? The cells are not larger than a newt.
39. 'to preserve' is teleology. The part in brackets does not make sense.
40. Where . . . the . . . disassembles. The last sentence is very inelegant.

Yes, in Figure 2, first sentence, 'has' should have been 'have'; subject 'roles'.

Index